SOFTWARE ENGINEERING TECHNIQUES APPLIED TO AGRICULTURAL SYSTEMS

An Object-Oriented and UML Approach

Applied Optimization

VOLUME 100

Series Editors:

Panos M. Pardalos
University of Florida, U.S.A.

Donald W. Hearn
University of Florida, U.S.A.

SOFTWARE ENGINEERING TECHNIQUES APPLIED TO AGRICULTURAL SYSTEMS

An Object-Oriented and UML Approach

By

PETRAQ J. PAPAJORGJI
University of Florida, Gainesville, Florida

PANOS M. PARDALOS
University of Florida, Gainesville, Florida

 Springer

Library of Congress Cataloging-in-Publication Data

Papajorgji, Petraq J.
   Software engineering techniques applied to agricultural systems : an object-oriented and UML approach / by Petraq J. Papajorgji, Panos M. Pardalos.
      p. cm. — (Applied optimization ; v. 100)
   Includes bibliographical references and index.
   ISBN 0-387-28171-1 (e-book)
     1. Agriculture—Data processing. 2. Software engineering. 3. Object-oriented programming (Computer science) 4. UML (Computer science) I. Pardalos, P.M. (Panos M.), 1954– II. Title. III. Series.

S494.5.D3P27 2006
630´.2´085—dc22

                                      2005051562

AMS Subject Classifications: 68N99, 68U35

	e-ISBN-10: 0-387-28171-1
ISBN-13: 978-1-4419-3926-5	e-ISBN-13: 978-0387-28171-1

© 2006 Springer Science+Business Media, Inc.
Softcover reprint of the hardcover 1st edition 2006

springeronline.com

To our children:

Dea Petraq Papajorgji

and

Miltiades Panos Pardalos

Contents

Preface

This book is an effort to bring the application of new technologies into the domain of agriculture. Historically, agriculture has been relatively behind the industrial sector in using and adapting to new technologies. One of the reasons for the technological gap between industrial and agricultural sectors could be the modest amounts of investments made in the field of agriculture compared to the impressive numbers and efforts the industrial sector invests in new technologies. Another reason could be the relatively slow process of updating the student's curriculum with new technologies in university departments that prepare our future specialists in the field of agriculture.

With this book, we would like to narrow the technological gap existing between agriculture and the industrial sector in the field of software engineering. We have tried to apply modern software engineering techniques in modeling agricultural systems. Our approach is based on using the object-oriented paradigm and the Unified Modeling Language (UML) to analyze, design, and implement agricultural systems.

Object-oriented has been the mainstream approach in the software industry for the last decade, but its acceptance by the community of agricultural modelers has been rather modest. There are a great number of researchers who still feel comfortable using traditional programming techniques in developing new models for agricultural systems. Although the use of the object-oriented paradigm will certainly not make the simulation models predict any better, it will surely increase the productivity, flexibility, reuse, and quality of the software produced.

The success of the object-oriented approach is mostly due to the ability of this paradigm to create adequate abstractions. Abstraction is an effective way to manage complexity as it allows for focusing on important, essential, or distinguishing aspects of a problem under study. Object-oriented is the best approach to mimic real world phenomena. Entities or concepts in a problem domain are conceived as objects provided with data and behavior to play a well-defined role. Objects can represent any thing in the real world, such as a person, a car, or a physiological process occurring in a plant. The use of objects enormously facilitates the process of conceptual modeling, which can be defined as the process of organizing our knowledge of an application domain into orderings of abstractions to obtain a better understanding of the problem under study. Conceptual modeling makes heavy use of abstraction, and the object-oriented approach, unlike other programming paradigms, provides direct support for the principle of abstraction.

Currently, UML is an industry standard for visualizing, specifying, constructing, and documenting all the steps of the software development. UML allows for presenting different views of the system under study using several diagrams focusing on the static and the dynamic aspects of the system. UML can be used in combination with a traditional programming environment, but its power and elegance fits naturally with the object-oriented approach.

One of the most beneficial advantages of UML is its ability to design a Platform Independent Model (PIM) that is a representation of the model using a high level of abstraction. Details of the model can be expressed clearly and precisely in UML as it does not use any particular formalism. The intellectual capital invested in the model is insulated from changes in the implementation technologies.

A Platform Specific Model (PSM) is developed by mapping a PIM to a particular computer platform and a specific programming environment. A mapping process allows the transformation of the abstract PIM into a particular PSM. This two-layer concept, a PIM and the corresponding PSM, keeps the business logic apart from the implementation technologies. Experience shows that the business logic has a much longer life than the implementation technologies. Changes and evolution of the implementation technologies should not have any impact on the business model.

The book is divided into two parts. Part one presents the basic concepts of the object-oriented approach, their UML notations, and an introduction to the UML modeling artifacts. Several diagrams are used to present the static and dynamic aspects of the system. There are an ample number of examples taken from the agriculture domain to explain the object-oriented concepts and the UML modeling artifacts. In this part of the book, a short introduction to design patterns explains the need for using proven solutions to agricultural problems.

Part two deals with applying the object-oriented concepts and UML modeling artifacts for solving practical and real problems. Detailed analyses are provided to show how to depict objects in a real problem domain and how to use advanced software engineering techniques to construct better software. Examples are illustrated using the Java programming language.

The book aims to present modeling issues a designer has to deal with during the process of developing software applications in agriculture. Although the Java programming language is used to illustrate code implementation, this book is not intended to teach how to program in Java. For this topic, we would recommend the reader to look for more specialized books. There is a chapter in this book that introduces the reader to some of the design patterns that we have used in agricultural applications. In no way do we pretend to have covered entirely the subject of how to use design patterns in software development. For an advanced and full presentation of the design patterns, we strongly suggest the reader to consider the well-known book *Design Patterns Elements of Reusable Object-Oriented Software* [GHJ95].

Our approach is based on the Rational Unified Process (RUP) methodology, although it does not rigorously follow this methodology. Our focus is on presenting modeling issues during the analysis and design of agricultural systems. For a more detailed and advanced approach to RUP, the reader needs to consult more specialized books.

What makes our book of unique value? Well, we have assembled in a comprehensive way a wide range of advanced software engineering techniques that will allow the reader to understand and apply these techniques in developing software applications in agriculture and related sciences. Agricultural systems tend to be more abstract than business systems. Everyone has a good understanding of how to use an ATM (Automated Teller Machine). The use of an ATM is a classic example, used in many publications, to explain what an object is and how to build a UML diagram. The process of photosynthesis or the interaction of a plant with the surrounding environment, just to name a few typical agricultural examples, is less known to a large number of readers. Modeling a plant as an object provided with data and behavior may not be as straightforward as modeling an ATM. Therefore, the book aims to provide examples and solutions to modeling agricultural systems using the object-oriented paradigm and the Unified Modeling Language.

The book is intended to be of use to anyone who is involved in software development projects in agriculture: managers, team leaders, developers, and modelers of agricultural systems. Developing a successful software project in agriculture requires a multidisciplinary team: specialists from different fields with different scientific backgrounds. It is crucial to the success of the project that specialists involved in the project have a common language that everybody understands. We find that UML is an excellent tool for analyzing, designing, and documenting software projects. Models can be developed visually and using plain English (and any other language for that matter) and can be understood by programmers and non-programmers alike. Thus, collaboration between these groups is substantially improved by increasing the number of specialists directly involved in the process of software design and implementation.

The book was written having always in mind the important number of specialists that still develop agricultural models using traditional approaches. There are ample step-by-step examples in this book that show how to depict concepts from a problem domain and represent them using objects and UML diagrams. We hope this book will be useful to these researchers and help them make a soft switch to the object-oriented paradigm. We hope readers will find this book of interest.

Gainesville, Florida, May 2005

PETRAQ PAPAJORGJI AND PANOS PARDALOS

Acknowledgments

A book is never the result of the author's work only, and this book does not make exception form this general rule. Several are they who deserve credit for their objective and unselfish help that made this book better. We would like to express our gratitude to all of them for their criticism and suggestions that contributed to improve the quality of our work.

Special thanks go to Dr. Tamara Shatar, a young and talented scientist who joined our team at IFAS, University of Florida, just after finishing her PhD from University of Sidney, Australia. Dr. Shatar was instrumental in developing the first UML diagrams representing various water-balance and irrigation-scheduling models.

We are very grateful to Alaine Margarete Guimaraes for her meticulous review of the manuscript. Alaine's help was crucial to finding most of the subtle discrepancies of the book.

Some instructive and insightful comments were made by Dr. Shrikant Jagtap, whose immense publishing experience has been very helpful not only for publishing this book but also for other work in the past. We thank Ray Buclin, Professor at the University of Florida, for the time he spent to read the book and make useful suggestions. The last but not certainly the least, Dr. Hugh Bigsby and Dr. Ram Ranjan deserve our gratitude for reading, one by one, all pages of the book and providing important insights that improved our work.

PART 1: CONCEPTS AND NOTATIONS

In the first part of the book, basic principles and concepts of the object-oriented paradigm as well as the corresponding UML notation are presented. There are a large number of examples that describe the basic object-oriented concepts and their UML notation. Chapters 1 to 7 belong to the first part of the book. The material presented in these chapters builds the basis for approaching the applications presented in the second part of the book.

Chapter 1

PROGRAMMING PARADIGMS

1. HISTORY OF INCREASING THE LEVEL OF ABSTRACTION

When developing software one deals with levels of abstraction, ranging from the real world where the problem represents the highest level of abstraction, to machine language that represents the solution in the lowest level of abstraction. Between the highest and lowest levels of abstraction, one should develop software in as many levels of abstraction as the problem demands.

The history of software development is the history of increasing the level of abstraction at each step. In the early days of computing, programmers used to represent programs using the lowest level of abstraction, by sending binary instructions corresponding to native CPU instructions to the computer to be executed. The main challenge programmers faced in those early days was the efficient use of the very limited amount of memory space available.

Later, an important software innovation took place: Assembly language was developed under the pressure of an increasing number of new and larger applications in the real world. Assembly language is based on abstractions designed to allow programmers to replace the 0s and 1s of native computer instructions by mnemonics. An assembler was used to translate the mnemonics into 0s and 1s corresponding to the native processor instructions, allowing the programmer to concentrate more on the problem than in programming error-prone details. Mnemonics were a higher level of abstraction. Writing code was less time-consuming and programs were less prone to error. The increased level of abstraction was followed by an increase in the level of productivity, software quality, and longevity.

The next big jump in using abstraction in code writing was the advent of the third-generation languages (3GLs). In these languages, a number of machine instructions needed to execute a certain operation like PRINT, for example, were grouped into macro instructions with a name. The code would be written using these macro instructions and a translator would translate macros into a sequence of 0s and 1s that now are called *machine code*. The high-level constructs used to write code allowed programmers to not be limited by the hardware capabilities. If a new hardware was used with a different set of instructions, then new translators were created to take into consideration new changes and generate code targeted to the new hardware. The ability to adjust code to different machines was referred to as *portability*.

The new set of tools programmers could use increased the number of domains and applications for which computer solutions were possible. So code was developed for each new and different problem. System vendors began to use 3GLs instead of assembly languages, even to define operating systems services. While 3GLs raised the level of abstraction of the programming environment, operating systems raised the level of abstraction of the computing platform.

Structured programming gave a boost to the use of control abstractions such as looping or if-then statements that were incorporated into high level programming languages. The control structures allowed programmers to abstract out certain conditions that would affect the flow of execution. In the structured programming paradigm, software was developed by first making an inventory of the tasks needed to be achieved. Then, each task was decomposed into smaller tasks until the level of the programming language statement was reached. During all the phases of analysis and design in structural programming, the focus was on how to refine step-by-step tasks that needed to be accomplished.

Later, abstract data types were introduced into the programming languages that would allow programmers to deal with data in an abstract way, without taking into consideration the specific form in which data were represented. Abstract data types hide the specific implementation of the structure of the data as their implementation was considered a low level detail. Programmers did not have to deal with this low level of detail; instead, they manipulated the abstract data types in an abstract way.

The demand for developing more complex systems, and in shorter time, made necessary a new, revolutionary way of looking at the software development: The object-oriented paradigm. According to this new paradigm, the basic building block is an object. The object-oriented approach tries to manage complexity by abstracting out knowledge from the problem domain and encapsulating it into objects. Designing software means depicting objects from the problem domain and providing them with specific responsibilities. Objects dialog with each other in order to make use of each other's capabilities. Functionality is achieved through dialog among objects.

In the object oriented paradigm, the analysis and design processes start with a more abstract focus. The main focus is to identify which operations need to be accomplished and who would accomplish these operations. The corresponding responsibilities are to be distributed to objects. Objects are to be provided with the necessary data and behavior in order to play the particular role they are assigned to. Each object knows its responsibilities and it is an active player. Rarely are objects created to stand by themselves outside any collaboration with other objects.

One of the most important recent achievements that represented a great breakthrough in software development is what we refer to as design patterns. Design patterns are descriptions of communicating objects and classes that are customized to solve a general design problem in a particular context [GHJ95]. As a collaboration, a pattern provides a set of abstractions whose structure and behavior work together to carry out some useful functions [BRJ99]. They present recurring solutions to software design problems that occur in real world application development. Design patterns abstract the collaboration between objects in a particular context and could be used and reused again and again. The use of design patterns in software engineering moved the level of abstraction higher, closer to the problem level and away from the machine language level.

For many years, software development companies have developed applications in a number of languages and operating systems. Isolated islands of applications developed in different programming environments and operating systems make it difficult to achieve a high level of integration that is demanded by the age of the Internet. In order to be competitive, companies are now forced to look for ways of building communication bridges between these isolated islands.

Object Management Group (OMG) was created in 1989 to develop, adopt, and promote standards for the development and deployment of applications in distributed heterogeneous environments [Vin97], [VD98]. OMG's response to this challenging problem was CORBA (Common Object Request Broker Architecture). CORBA enables natural interoperability regardless of platform, operating system, programming language, and even of network hardware and software. With CORBA, systems developed in different implementation languages and operating systems do not have to be rewritten in order to communicate. By raising the level of abstraction above the implementation languages and operating systems, CORBA made a tangible contribution to the longevity of the software.

Another important event that significantly influenced the software engineering world was the use of visual modeling tools. Modeling is a well-known engineering discipline as it helps one to understand reality. Models of complex systems are built because it is difficult to understand such a system in its entirety. Models are needed to express the structure and the behavior of complex systems. Using models makes it possible to visualize and control system's architecture.

In the early 90s, there were several modeling languages used by the software engineering community. The most well-known methodologies were Booch, Jacobson's (Object-Oriented Software Engineering) and Rumbaugh's (Object Modeling Technique). Other important methods were Fusion [CAB94], Shllaer-Mellor [ShM88], and Coad-Yourdon [CY91]. All these methods had strengths and weaknesses. An important event occurred in the mid-90s when Booch, Jacobson, and Rumbaugh began adopting ideas from each other that led to the creation of the Unified Modeling Language or as it is known best, the UML. UML is a standard language for visualizing, specifying, constructing, and documenting object-oriented systems [BRJ99].

UML uses a set of graphical symbols to abstract things and their relationships in a problem domain. Several types of diagrams are created to show different aspects of the problem. Models created using UML are semantically richer than the ones expressed in any current object-oriented language and they can express syntax independently of any programming language. When a UML model is translated into a particular programming language, there is loss of information. A UML model is easy to read and interpret as it is expressed in plain English. Therefore, formal models raise the programming abstraction level above the 3GLs in a profound way.

In 2002, OMG introduced a new and very promising approach to software development referred to as the Model Driven Architecture approach, known as the MDA. MDA is about using modeling languages as programming languages [Fra03].

Most commonly, software models are considered to be design tools while code written in programming languages are considered to be development artifacts. In most of the software development teams, the role of the designer is quite separated from the role of the developer. A direct consequence of this separation is the fact that design models often are informal, and are used by developers only as guidelines for software development. This separation of roles is the common source of discrepancies that exist between design models and code.

MDA aims to narrow the gap existing between the designer and the developer by providing for producing models that can be compiled and executed in the same environment. Therefore, models will not be only a design artifact, but an integral part of the software production process. The MDA approach to software development is based on building platform independent models (PIM) that later can be mapped into platform specific models (PSM) that take into consideration specific implementation issues, such as platforms, middleware, etc. Specific models are then used by code generators to automatically create the implementation code.

The MDA approach has already been applied to a variety of computing technologies such as Sun's J2EE and EJB, Microsoft's .NET, XML, OMG's CORBA, and an increasing number of software companies are adopting this approach. Although it is quite early to evaluate the impact MDA is having in the software industry, the future looks promising. By narrowing the gap between designers and developers, MDA considerably raised the level of abstraction in software development.

2. OBJECT-ORIENTED VERSUS OTHER PROGRAMMING PARADIGMS

The most common ways to approach modeling of a software problem are the following: From an algorithmic perspective and from an object-oriented perspective.

The first approach is represented by structured programming, known nowadays as traditional programming. Structured programming is characterized by the use of the top-down approach in design and software construction. According to this paradigm every system is characterized, at the most abstract level, by its main function [Mey88]. Later, through an iterative process, the top function is decomposed into simpler tasks until the level of abstraction provided by a programming language is reached. Each of the steps of the iteration can be considered as a transformation process of the input task into a more suitable one required by the next iteration step. The main building block is the function or procedures. The expected behavior of a system is represented by a set of functions or procedures.

Data flow diagrams are used to represent a functional decomposition. A data flow diagram can be considered as a graph, with nodes representing the data transformations that occur at each step of the iteration. Structural programming is data-centric; functional components are closely related and dependent to the selected data structure. Therefore, changes to the data representation or the structure of the algorithm used may have unpredictable results. The entire data model needs to be considered when designing functionalities for the system. Changes to the data structure or to procedures, usually affect the entire system.

As data are the center of the traditional programming approach, the analysis and design phases of software construction deal with what is referred to as the *data model*. A data model is conceived as a flat structure, viewed from above. Data are distributed into tables and relationships between data are represented by relationships between tables. Relationships are implicit, using the foreign keys. Two tables are related among them if both have at the least a field in common, that contains the same type of data. Therefore, relationships in a data model are bidirectional.

The reasons why the structured programming approach is not a well-suited approach for software development has been widely discussed in the software engineering literature [Mey88], [Boo86], to name a few. Although the list of disadvantages of using the structured programming paradigm is rather long, we would like to point out one particular handicap of this approach that has important consequences in modeling agricultural systems: Its lack of support for concurrency. There are many agricultural systems that coexist and interchange information between them. Intercropping, one of most relevant examples that use concurrency, is the process when two or more crops share the same natural resources (i.e., water, soil, and weather). In order to simulate the growth of one of the crops, other participating crops must be considered at the same time, as they compete for the same resources.

It is difficult to develop a software system that allows for concurrency using traditional programming languages such as FORTRAN. Complex and difficult-to-use data structures need to be created to allow for developing an intercropping system. The number of crops simultaneously used by the system has to be known in advance to be adequately represented in the data structures. Furthermore, transforming an existing system that does not allow intercropping into a system that allows for it would certainly require the reexamination of the entire existing system.

The software industry has embraced, for more than a decade, a more revolutionary approach to software development, the object-oriented paradigm. [Mey88] defines the object-oriented software construction as: *Object-oriented construction is the software development method which bases the architecture of any software system on modules deduced from the types of objects it manipulates (rather than the function or functions that the system is intended to ensure).* The modularity in the object-oriented approach is the class.

The focus of this new programming paradigm is the class that directly represents concepts (abstract or concrete) of a particular domain to which the software applies. Objects created from classes are the best way to mimic real-life problems; an object can represent an abstract concept, such as a photosynthesis process or a real concept such as plant, soil, or weather. Objects are provided with data and behavior and interact with each other. Functionality is achieved through dialog amongst objects. The dialog between objects is represented by relationships. Two objects are related if they need to access data and behavior from the other object. Relationships between objects are represented by object diagrams in an *object model*. The activity of describing relationships between objects in a problem domain is referred to as *object modeling*.

The power of the object-oriented approach as modeling tool is the subject of the following chapters of this book.

Chapter 2

BASIC PRINCIPLES OF THE OBJECT-ORIENTED PARADIGM

1. ABSTRACTION

One of the most appreciated advantages of object-oriented versus other modern programming paradigms is the direct support for each of the most important and used principles of abstraction. The Dictionary of the Object Technology defines abstraction as: "Any model that includes the most important, essential, or distinguishing aspects of something while suppressing or ignoring less important, immaterial, or diversionary details. The result of removing distinctions so as to emphasize commonalities." Abstraction is an effective way to manage complexity, as it allows for concentrating on relevant characteristics of a problem. Abstraction is a very relative notion; it is domain and perspective dependent. The same characteristics can be relevant in a particular context and irrelevant in another one.

The abstraction principles used in the object-oriented approach are: Classification/instantiation, aggregation/decomposition, generalization/ specialization and grouping/individualization. By providing support for the abstraction principles, the object-oriented paradigm makes it possible to use conceptual modeling as an efficient tool during the phases of analysis and design. Conceptual modeling can be defined as the process of organizing our knowledge of an application domain into hierarchical rankings or orderings of abstraction, in order to better understand the problem in study [Tai96].

Classification is considered to be the most important abstraction principle. It consists of depicting from the problem domain things that have similarities and grouping them into categories or classes. Things that fall into a class/category have in common properties that do not change over time. *Instantiation* is the reverse operation of *classification*. It consists of creating individual instances that will fulfill the descriptions of their categories or classes. The majority of object-oriented languages provide capabilities for creating instances of classes/categories.

Figure 2-1 shows an example of *classification* and *instantiation*. Concept *Tractor* represents a set of properties that are typical for a tractor, regardless of their brand, horsepower, etc. Therefore, concept *Tractor* represents a classification. *Bob's Tractor* is a particular tractor that has some particular properties, the most important being that it is Bob's property. Therefore, concept *Bob's Tractor* represents an instantiation.

Classification and Instantiation

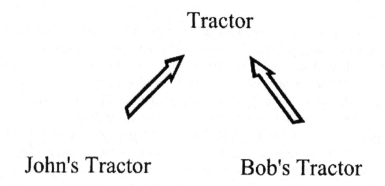

Figure 2-1. Examples of classification and instantiation.

The second abstraction principle is *aggregation*. *Aggregation* refers to the principle that considers things in terms of part-whole hierarchies. Concepts in a problem domain can be treated as aggregates (i.e., composed of other concepts/parts). A part itself can be considered as composed of other parts of smaller granularity. *Decomposition* is the reverse operation of *aggregation*; it consists of identifying parts of an aggregation. Object-oriented languages provide support for aggregation/decomposition by allowing objects to have attributes that are objects themselves. Thus, complex structures can be obtained by using the principle of *aggregation.* Note that some authors use the term *composition* instead of *aggregation.*

Figure 2-2 shows an example of *aggregation* and *decomposition.* Concept *Tractor* can be considered as an aggregation/composition of other concepts such as *Chassis, Body,* and *Engine.* Concept *Body* can be considered as one of the parts composing a more complex concept such as *Tractor.*

Aggregation and Decomposition

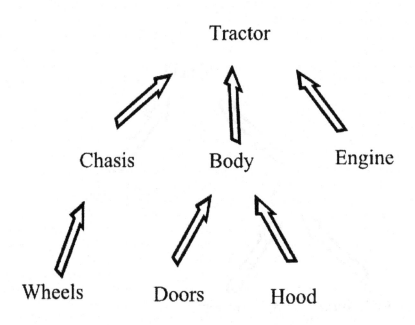

Figure 2-2. Example of aggregation and decomposition.

The third abstraction principle is *generalization. Generalization* refers to the principle that considers construction of concepts by generalizing similarities existing in other concepts in the problem domain. Based on one or more given classes, *generalization* provides the description of more general classes that capture the common similarities of given classes. *Specialization* is the reverse operation of *generalization.* A concept A is a specialization of another concept B if A is similar to B and A provides some additional properties not defined in B.

Object-oriented languages provide support for generalization/ specialization as they allow for creating subclasses of exiting classes and/or creating more general classes (superclasses) of existing classes. Creating a subclass of an existing class corresponds to specialization and creating a superclass of an existing class corresponds to generalization. It is important to note that concept A is a *generalization* of concept B if and only if B is a *specialization* of concept A [Ped89]. Figure 2-3 shows an example of *generalization* and *specialization*.

Generalization and Specialization

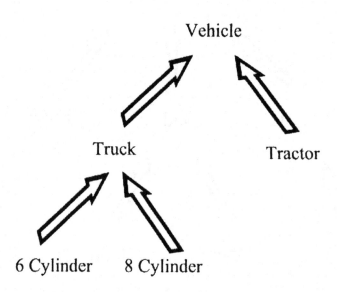

Figure 2-3. Example of generalization and specialization.

Concept *Truck* is a specialization of concept *Vehicle.* This is because *Truck* has all the properties of concept *Vehicle* and some additional ones that make it a special *Vehicle.* In reverse, concept *Vehicle* is a *generalization* of concept *Truck,* as all trucks are vehicles.

The fourth abstraction and perhaps the least obvious, is *grouping* [Tai96]. In conceptual modeling, often a group of concepts needs to be considered as a whole, not because they have similarities but because it is important that they be together for different reasons. Object-oriented languages provide a mechanism for grouping concepts together such as sets, bags, lists, and dictionaries. *Individualization* is the reverse operation of *grouping*. It consists of identifying an individual concept selected among other concepts in a group. Individualization is not as well established as a form of abstraction [Tai96]. Figure 2-4 shows an example of *grouping* and *individualization*.

Grouping and Individualization

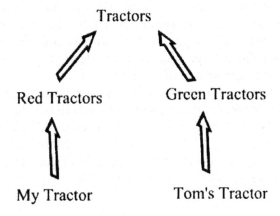

Figure 2-4. Example of grouping and individualization.

All tractors used in a farm can be grouped in one category regardless of their brand, color, horsepower, and year of production, and be represented by one concept such as *Tractors*. In case we need to use one of them with a certain horsepower, then we need to browse the set of tractors and find that particular individual that satisfies our needs. In this case, we have individualized one element of the set based on some particular criterion. When we say *Tom's Tractor*, we have used the ownership as criterion for individualizing one of the tractors, the one that belongs to Tom.

2. ENCAPSULATION

The Dictionary of the Object Technology defines encapsulation as: "The physical location of features (properties, behaviors) into a single black box abstraction that hides their implementation behind a public interface."

Often, encapsulation is referred to as "information hiding." An object "hides" the implementation of its behavior behind its interface or its "public face." Other objects can use its behavior without having detailed knowledge of its implementation. Objects know only the kind of operations they can request other objects to perform. This allows software designers to abstract from irrelevant details and concentrate on what objects will perform.

An important advantage of encapsulation is the elimination of direct dependencies on a particular implementation of an object's behavior. The object is known from its interface and clients can use the object's behavior by only having knowledge of its interface. The particular implementation of an object's interface is not important. Therefore, the implementation of the object's behavior can change any time without affecting the object's use. Encapsulation helps manage complexity by identifying a coherent part of this complexity and assigning it to individual objects.

The fact that an object hides the implementation of its behavior by exposing only its "public face" could be beneficial to other objects that need its behavior. The "interested" objects could consider more than one option while looking for a specific functionality that satisfies their needs. They need only to "examine" the interfaces of candidate objects. Objects with similar behavior could serve as substitutes to each other.

3.   MODULARITY

The Dictionary of the Object Technology defines modularity as: "The logical and physical decomposition of things (e.g., responsibilities and software) into small, simple groupings (e.g., requirements and classes, respectively), which increase the achievements of software-engineering goals."

Modularity is another way of managing complexity by dividing large and complex systems into smaller and manageable pieces. A software designing method is modular if it allows designers produce software systems by using independent elements connected by a coherent, simple structure. [Mey88] defines a software construction method to be modular if it satisfies the five criteria:

Modular Decomposability; a software construction method satisfies Modular Decomposability if it helps in the task of decomposing a software problem into a small number of less complex sub-problems, connected by a simple structure, and independent enough to allow further work to proceed separately on each of them.

Modular Composability; a software construction method satisfies Modular Composability if it favors the production of software elements which may then be freely combined with each other to produce new systems, possible in an environment quite different from the one in which they were initially developed.

Modular Understandability; a software construction method satisfies Modular Understandability if it helps produce software in which each module can be understood without having to examine other interrelated modules.

Modular Continuity; a software construction method satisfies Modular Continuity if a small change in the requirements of will impact just one or a small number of modules.

Modular Protection; a software construction method satisfies Modular Protection if the effect of an exception occurring at runtime will impact only the corresponding module or a few neighboring modules.

The concept of Modularity and the principles for developing modular software in the object-oriented approach are encapsulated in the concept of class. Classes are the building blocks in the object-oriented paradigm.

Chapter 3

OBJECT-ORIENTED CONCEPTS AND THEIR UML NOTATION

1.   OBJECT

Booch [BRJ99] defines an object as "a concrete manifestation of an abstraction; an entity with a well-defined boundary and identity that encapsulates state and behavior; an instance of a class."

In other words, an object is a concept, abstraction, or thing with well-defined boundaries and meaning in the context of a certain application. For example, in the domain of crop simulation models, *Plant* can be an object as it is an abstraction of different plants that represents most of their main characteristics. In the same manner, *Soil* can be an object as it is an abstraction that represents what is common to many types of soil.

An object represents an entity that can be physical or conceptual. Object *Plant* represents a physical entity; we can see a plant with its root system, leaves, and stems. In the same way, object *Soil* represents a physical entity as we can see soil surface and its composing layers if we are looking at a soil slope.

Often in crop simulation models an abstract entity is used, named *SoilPlantAtmosphere* to represent features and data that pertain to soil, plant, and atmosphere all together. The object *SoilPlantAtmosphere* represents an entity that is conceptual; it is artificially created to represent data and behavior that are needed in the simulation process that cannot be assigned to only one of the above-mentioned entities.

The problem domain is crucial while selecting entities to be future objects. Objects named with the same name may have a very different meaning in the context of different domains. For example, object *Tomato* in the crop simulation domain may represent a different concept from the object *Tomato* created in the supermarket domain. In the first case, object *Tomato* is an abstraction for studying the effects of soil and weather in the crop growth while in the second case, object *Tomato* is an abstraction used to study the situation of the market for a certain period of time.

Objects encapsulate state and behavior. The state of an object is one of the many possible conditions in which the object may exist. Object *Plant* can be in different conditions during a simulation. Initially, *Plant* exists in the form of the seed, later on in the stage of vegetation, flowering, and finally, in the stage of maturity.

The state of an object is provided by a set of properties or attributes, with their values. As objects represent dynamic entities, their state may change over time (i.e., the values of attributes may change over time). Thus, the attribute *growthStage* of object *Plant* will have the value "in vegetation" in the early days of the simulation and the value "in maturity" in the late days of the simulation.

The behavior of the object is defined by its role in the dialog with other objects. In other words, the behavior of an object represents how the object acts and reacts to the requests of other objects. Object's behavior is represented by a set of messages that the object can respond to. Object *Weather* should be able to provide solar radiation data to other objects requesting it; therefore, it responds to the message *getSolarRadiation*. Providing weather data such as minimum and maximum temperature, wind speed, rainfall, is part of the behavior of object *Weather*.

Every object has a unique identity. Identity makes objects distinguishable, even in cases where they have the same attributes. The identity of the object does not depend on the values of its attributes. For example, the identity of the object *Plant* does not change, although some of its attributes, such as *growthStage,* may change over time. Each time an object is created, a unique identity is provided to it that will identify the object for the rest of its life.

2.   CLASSES

Booch [BRJ99] defines a class as description of a set of objects that share the same attributes, operations, relationships, and semantics. Classes are the most important building block of any object-oriented system. In a specific domain, there are many objects that can share commonalities. It is important to abstract features that are common to objects. Then, these common features will be used to construct a class. Abstraction is used to depict commonalities between objects and construct classes. A class is not an individual object, but rather a pattern to create objects with same properties and behavior. Objects created by a class are instances of this class.

Some authors refer to a class as an "object factory," a factory that knows how to produce objects. Other metaphors are "rubber stamps" or "blueprints."

A class has a unique name and all classes defined in a domain have different names.

Figure 3-1. Examples of objects.

In Figure 3-1 there are three objects: A pear, a strawberry, and grapes. These objects have different colors, shapes, and taste but they have in common the fact that they are fruits. They all can be represented by class *Fruits* and each of them is an instance of the class *Fruits*. The class *Fruits* should be designed to represent common characteristics of each of the instances.

3.   ATTRIBUTES

An attribute is a named property of a class that describes a range of values that instances of the property may hold [BRJ99]. Attributes hold information about the class and they define the state of the object created by the class. Each attribute can hold values independently of one another. A class may have any number of attributes or no attribute at all.

An attribute has a name, and it is advisable to name attributes with meaningful names that represent this particular abstraction expressed in the attribute. An attribute has a type that defines the kind of values that can be stored in it. An attribute is an abstraction of the kind of the data an object of the class may have as value.

Class *Fruits* has an attribute named *fruitName* used to hold the name of the fruit. All instances of this class will have this attribute, but the value of the attribute may be different. If an instance of the class *Fruit* is created and the attribute *fruitName* is set to "Apple," then it shows that the instance created is an apple.

4.   OPERATIONS

An operation is the implementation of a service that can be requested from any object of the class to affect behavior [BRJ99]. The set of all operations of a class define its behavior. The behavior of the class is defined during the analysis and design phases, and depends on the role that the class has in the domain.

A class may have any number of operations but the number of operations should reflect the behavior of the class. An operation has a name that usually is a verb or a verb phrase that represents some particular behavior of its class.

An operation has its signature that is the name and the type of all parameters used by the operation. In cases when the operation returns a result, the type of the returned value should be specified. Figure 3-2 shows the UML symbol for class *Soil*.

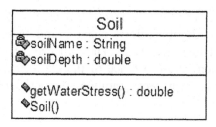

Figure 3-2. Example of a class.

In UML, a class is represented by a rectangular icon divided into three compartments as shown in Figure 3-2. The top-most compartment contains the name of the class. The name of the class starts with a capital letter. The middle compartment contains the list of attributes of the class. The name of an attribute starts with a lowercase letter. Attributes have a *type*; attribute *soilName* is of the type *String*. This compartment is considered the "data" compartment, as the attributes hold data. The bottom compartment lists the operations or the methods of the class. A method represents a specific behavior the class provides. In the case of class *Soil*, the method *getWaterStress()* calculates the water stress for this particular type of soil. If the method returns a result, the type of the result is defined. In some cases, operations require parameters. In this case, the type of the parameter should be defined. In the case of class *Soil*, the method *getWaterStress()* returns a result of the type *double* and does not require any parameters. Figure 3-3 shows the implementation in Java of the class *Soil* defined in Figure 3-2.

```
1    package Soil;
2
3    public class Soil {
4
5        private String soilName;
6        private double soilDepth;
7        public Soil() {}
8        public double getWaterStress(){
9            double waterStress = 0.0;
10           //here goes the body of the method, to be implemented
11           return waterStress;
12       }
13   }
```

Figure 3-3. Java definition of class Soil.

Line 1 in Figure 3-3 shows the package or the subdirectory where the Java code is stored. The concept of package will be introduced later in this book. Line 3 shows that a new class, referred to as *Soil*, is defined. Lines 5 and 6 show the definition of attributes for class *Soil*. Attribute *soilName* is of type String; the values this attribute can hold should be of type String. Attribute *soilDepth* is of type *double*, the values this attribute holds should be of type *double*. Line 7 describes the default constructor for class *Soil*. A constructor is the mechanism in Java that creates instances of a class. Lines 8 through 12 are the definition of the operation *getWaterStress*. Line 8 shows that the result of the operation is of type *double*. Line 9 defines a local attribute named *waterStress* of type *double*. This attribute will hold the calculated value of water stress parameter. Line 10 is a comment in Java that shows that the logic for water stress calculation needs to be provided by the user. Line 11 returns the value of the calculated parameter to the object that asked for it.

The number of attributes and operations that a class is provided with directly affects the behavior of the class. Designing the attributes and the operations of a class is not an easy task. It has to do with the role and the responsibilities the class will have in the domain in study. It is through the attributes and the operations that the responsibilities of a class are carried out.

5.   POLYMORPHISM

Polymorphism is one of the most important features offered by the object-oriented paradigm. Polymorphism comes from the Greek term "polymorphos" meaning "having many forms." In object-oriented programming, it refers to the ability of the language to process object differently depending on their class (Webopedia at **http://www.webopedia.com**).

Szyperski [Szy99] defines polymorphism as "the ability of something to appear in multiple forms, depending on the context; the ability of different things to appear the same in a certain context." Essentiality the concept of polymorphism has to do with substitutability between objects; in certain contexts, one object could be substituted with another.

In order to better understand the concept of polymorphism, another related concept should be introduced: The concept of interface.

6.   INTERFACES

Let us introduce the concept of the interface using an example. Suppose that a class needs to be designed with the task of providing weather data for a crop simulation scenario. Weather data can be provided in different ways. Some authors [LKN02] have designed their classes to obtain weather data from a network of real-time weather stations. Others [HWH01] obtain weather data from text files saved locally in the system. It is desirable to design a system flexible enough to provide weather data from several sources. The fact that weather data could be provided in different ways does not affect the logic used to handle these data for calculating photosynthesis processes or water movement in soil. The scientific equations used to calculate photosynthesis will deliver the same results regardless of the source used to obtain the data: Be they read from text files, from a database management system, or obtained directly from an on-line weather station.

Most of the existing systems are designed to use one single source of data. The IBSNAT group has selected to develop a special format of text files to store the weather data [HWH01]. This way of forcing the system to a very specific way of obtaining weather data, limits other valuable sources of weather data to be used such as the ones provided by the on-line weather stations or other important sources.

As previously mentioned, the scientific calculations are not affected by the way weather data are obtained. Therefore, it will be beneficial to express the weather data used in a general way, independently of the particular data source. The object *Weather* should be designed in such a way that it would provide the data regardless of the specific source used. This object should only logically show the kind of data needed and ignore any particular way of obtaining the weather data. An object that plays this role is called an interface.

[BRJ99] defines an interface as "a collection of operations that are used to specify a service of a class or a component." [Szy99] defines an interface as "a contract between a client of the interface and a provider of an implementation of the interface." Figure 3-4 shows the UML symbols for an interface.

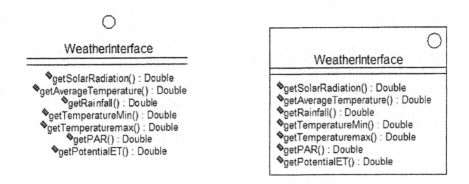

Figure 3-4. UML symbols for an interface.

As shown in Figure 3-4, each of the symbols can be used as they represent the same thing, the graphic symbol of an interface. Figure 3-5 shows an example of interface defined in Java.

```
1   package Weather;
2   public interface WeatherInterface {
3       public double getSolarRadiation();
4       public double getAverageTemperature();
5       public double getRainfall();
6       public double getTemperatureMin();
7       public double getTemperatureMax();
8       public double getPar();
9       public double getPotentialET();
10  }
```

Figure 3-5. Examples of interface definition in Java.

An interface only defines the kind of services an object should provide for use by other objects. It represents the set of messages that can be sent to an object created by a class that implements this interface. The implementing object will respond to any of the messages defined in the interface. The interface defines only the operations' signature composed of the operation name and the parameters required for its execution that must be conveyed in a message. There is no body in the methods or operations defined in the interface. The body for each of the methods will be provided by the class that implements this interface. Figure 3-6 shows an example of a class implementing an interface.

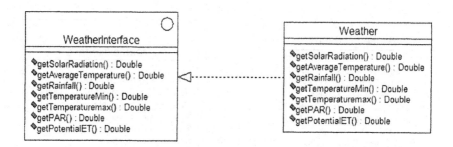

Figure 3-6. Class implementing an interface.

As shown in Figure 3-6, class *Weather* implements interface *WeatherInterface*. This contractual agreement, establishes obligations in both sides: The interface defines what classes should at the least implement as functionality when binding the interface, and the classes realize that the interface definition is what they are expected to implement at the least. Therefore, the behavior of class *Weather* would include at the least the behavior defined by the interface that the class is implementing. Class *Weather* may have other methods that are not related to *WeatherInterface*. An interface can be implemented by many classes; each of them would provide a particular implementation of the behavior defined by the interface. At the same time, a class may implement many interfaces, in which case the behavior of the class will include all the behaviors defined by the interfaces that the class implements. Figure 3-7 shows the example of class *Weather* implementing interface *WeatherInterface*.

```
1   package Weather;
2   public class Weather implements WeatherInterface {
3   private double solarRadiation;
4   private double rainFall;
5   private double potentialET;
6   private double temperatureMax;
7   private double temperatureMin;
8   private double par;
9   public double getSolarRadiation() {return solarRadiation;}
```

Figure 3-7. Class Weather implementing WeatherInterface (Part 1 of 2).

```
10  public double getRainFall() {return rainFall;}
11  public double getTemperatureMax() {return temperatureMax;}
12  public double getTemperatureMin() {return temperatureMin;}
13  public double getAverageTemperature() {
14     return (getTemperatureMin()+getTemperatureMax())/2;
15  }
16  public double par(){return par;}
17  public double getPotentialET() {return potentialET;}
18  public Weather() {}
19  }
```

Figure 3-7. Class Weather implementing WeatherInterface (Part 2 of 2).

As shown in Figure 3-7, the definition of class *Weather* includes all the operations defined in the interface *WeatherInterface*. The interface defines only the signature for the operation, not the body, whereas the class definition includes the signature's operation and the code for implementing it. As an example, line 13 in Figure 3-7 is the same as line 4 in Figure 3-5, as they both define operation's signature. In Figure 3-7, line 14 provides the code implementation for the operation.

A better approach for the problem defined at the beginning of this section is to design an interface that would define the functionalities that classes implementing the interface should provide, with each of the classes providing a particular solution to the problem of obtaining the weather data. Figure 3-8 shows two classes implementing the same *WeatherInterface*.

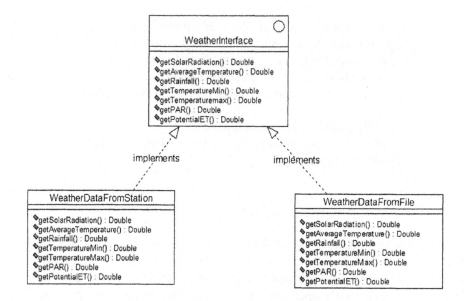

Figure 3-8. Example of two different classes implementing the same interface.

As shown in Figure 3-8, classes *WeatherDataFromStation* and *WeatherDataFromFile* implement the same interface *WeatherInterface*. Both classes have agreed to provide behavior for the functionality defined in the interface. Class *WeatherDataFromStation* will provide the behavior necessary to obtain weather data from an online station. As such, this class is responsible for providing the logic for solving issues such as connecting to the station, making transactions, and downloading the data. Class *WeatherDataFromFile* will provide the behavior for reading the weather data from a text file. These two classes provide different behavior to implement the same functionality that is obtaining weather data. Different behaviors provide the same results, but in different ways. The same functionality is provided in many forms, and this is the meaning of polymorphism.

Interfaces are an elegant way of defining polymorphism without involving any implementation. Two classes provide polymorphic behavior if they implement the same set of interfaces. Interfaces formalize polymorphism.

The basic concept behind polymorphism is substitutability. Classes implementing the same interface can be a substitute to each other; any one of them can be used instead of the other. Therefore, interfaces are the heart of the "plug and play" architecture. Classes, components, or subsystems can substitute for each other provided that they implement the same set of interfaces. Interfaces are a key concept while developing component-based environmental and agricultural software [PBB04]. Software designs that do not use interfaces are rigid, and difficult to maintain and reuse. Interfaces allow objects to communicate with each other without demanding detailed knowledge of object's internal logic and implementation.

7. COMPONENTS

There are several definitions of the component-based approach. [BRJ99] define a component as a "physical and replaceable part of a system that conforms to and provides the realization of a set of interfaces." This definition is broad and considers the component to be an organizational concept representing a set of functionalities that can be reused as a unit. According to this definition documents, executables, files, libraries and tables could be considered as components. The emphasis in this definition is on reuse.

Microsoft Corp. has a slightly different definition. In "Component Definition Model," [BBC99] define a component as a "software package which offers services through interfaces." The emphasis in this definition is on service provider. The service provider approach considers a component to be a piece of software that provides a set of services to its users. Services are grouped into coherent, contractual units referred to as interfaces [Bro99]. Users can utilize services offered by knowing the component's interface specification, or contract.

Both the reuse and service provider perspectives of a component introduce the important distinction between its public interface (what the component does) and its implementation in a particular programming environment (how the component does it). This distinction addresses the issue of how the component should be designed in order to be an independent and replaceable piece of software with minimal impact on the users.

In other words, components are reusable pieces of software code that serve as building blocks within an application framework. [BW98] conclude that although components fit better with the object-oriented technology, they are often used in non object-oriented programming environments representing a chunk of functionalities that a system should provide. Most such environments provide an ad hoc way of defining an interface, limiting its scope and use. The object-oriented paradigm provides formal languages for defining an interface and its contract, broadening its scope and reutilization opportunities.

In traditional programming, most of the times the focus is developing a single stand-alone system, where all variables and procedures are located in a single container, referred to as the main program. In contrast, the component-based approach has as its focus the building of a set of reusable components from which a family of applications can be assembled [Bro99]. The component-based paradigm addresses a number of important questions related to the optimal size of a component, the kind of documentation that needs to be provided so others can use components, and how the assembly of existing components needs to be performed.

Every component has a name that distinguishes it from other components. The name is a textual string. The UML symbol for a component is shown in Figure 3-9.

Figure 3-9. The UML symbol for a component.

There are three types of software components [BRJ99].

Deployment Components

Deployment components include components that can form an executable system, such as dynamic libraries (DLLs) and executables (EXEs). Included in this category are other object models such as COM+, CORBA and Enterprise Java Beans or object models such as dynamic Web pages and database tables.

Work Product Components

Work product components are created from source code files or data files. They do not directly participate in an executable system but are the work product of development that is used to create an executable system.

Execution Components

These components are created as a consequence of an executing system, such as COM+ object, which is instantiated from a DLL.

There is a great similarity between the concepts of component and class. The most important one is that both components and classes may implement a set of interfaces. They both can be used as modeling artifacts, i.e., participate in relationships, have dependencies, etc. [BRJ99].

The fundamental difference between components and classes is that a component usually offers its services only through interfaces, whereas a class may or may not implement any interfaces. A component is designed to be part of a system; it is a physical entity that offers its services through interfaces. A component may be reused across many systems. The process of creating components as modeling entities uses the principle of encapsulation.

8.   PACKAGES

[BRJ99] defines a package as a "general purpose mechanism for organizing elements into groups." Packages are used as containers to include modeling elements that are logically related and can be manipulated as a group. Modeling elements that can be included in a package may be classes, interfaces, components, different kinds of diagrams, and even other packages. A package does not represent any abstraction of the elements it owns. The ownership relation in a package is strong; if the package is destroyed, its contained elements are destroyed as well. A package has a name to identify it from other packages. The UML symbol for a package is shown in Figure 3-10.

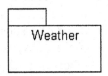

Figure 3-10. The UML symbol for a package.

Package *Weather* will contain all model elements related to weather. They can be classes, interfaces, components, and different diagrams. As packages are only used for storing purposes, they do not have any representation in the implementation, except for maybe as directories. During the process of creating packages, the principle of modularity is used.

9.   SYSTEMS AND SUBSYSTEMS

Let us suppose that all the activities in a farm need to be automated. A software system will be developed to cover the activities such as accounting, sales, inventory, and so on.

All relevant farm activities will be presented in a system. [BRJ99] defines a system as "a set of elements organized to accomplish a purpose and described by a set of models, possible from different viewpoints." A system possibly may be decomposed into a set of subsystems and should represent only the most relevant elements of the domain under study.

The UML symbol for a system is shown in Figure 3-11.

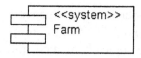

Figure 3-11. The UML symbol for a system.

[BRJ99] defines a subsystem as "a grouping of elements of which some constitute a specification of the behavior offered by other contained elements." A subsystem may contain classes, interfaces, components and other subsystems. A subsystem is a combination of a package and a class. As packages, subsystems have semantics that allow them to contain other model elements. Like classes, subsystems realize or implement a set of interfaces that give them behavior. The behavior of a subsystem is provided by classes and other subsystems included in the subsystem. The UML symbol for a subsystem is a combination of symbols of package and interface as shown in Figure 3-12.

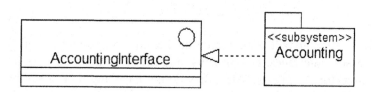

Figure 3-12. Example of UML symbol of a subsystem.

In Figure 3-12, the *Accounting* subsystem represents an implementation of operations defined in the interface *AccountingInterface*. Accounting, as part of activities in a farm, is quite independent. It collaborates intensively with other subsystems as it keeps track of all the expenses that occur in the farm. Accounting subsystem receives data from other subsystems and provides different financial reports.

When representing a particular problem domain as a system, decisions have to be made on how the system will be divided into subsystems and what classes will be included in each of the subsystems. Issues, such as the kind of behavior each subsystem should encapsulate and its size, are to be discussed and solved. Subsystems should be nearly independent with a well-defined purpose. They will be interacting with each other to provide the functionalities required by the system. Each subsystem encapsulates specific behavior and the behavior of the entire system will be provided by the set of created subsystems. Different levels of abstractions can produce different kinds of subsystems. A subsystem at one level of abstraction can be a system in another level of abstraction. In other words, a subsystem is only a composing part of a bigger and complex system. Encapsulation and modularity are the two principles of object-orientation that need to be considered during the phase of system analysis.

A subsystem should not expose any of its composing parts. No class in a subsystem should be visible outside the subsystem. No element outside the subsystem should have direct dependency on elements inside the subsystem. A subsystem should depend only on the interfaces of other model elements and other model elements will depend only on the set of interfaces of a subsystem. This way, a subsystem can be replaced by another one provided they implement the same set of interfaces.

As previously mentioned, the behavior of a subsystem is defined by the set of interfaces the subsystem implements. Subsystem's behavior will be provided only by classes or modeling elements that are included into the subsystem. There should not be any reference to any class outside the subsystem.

Figure 3-13 shows an example of decomposition of a farm system into subsystems. *Accounting* subsystem will include all the classes needed to carry out accounting responsibilities. The responsibilities of the subsystem are defined in the interface *IAccounting*. The subsystem *Production* includes all the classes representing operations occurring in a farm. The responsibilities of this subsystem are defined in the interface *IProduction*. The same way, the subsystem *Research&Development* will include all the classes defined to represent entities related to the research and development process. The responsibilities of this subsystem are defined in the interface *IResearch&Development*.

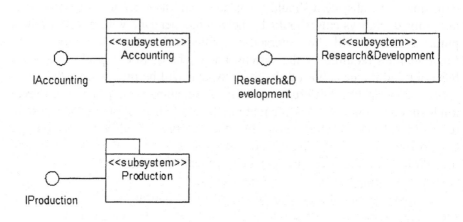

Figure 3-13. Decomposition of a system into subsystems.

There are some similarities between the concept of subsystem and the one of component. They both encapsulate a partial behavior of a bigger system and are designed to dialog with others to provide the functionality required by the system. Their behavior is defined by a set of interfaces for which they provide a polymorphic implementation. They both provide substitutable behavior; both can be replaced by other component/subsystems provided they realize the same set of interfaces. They are constructed using the same object-oriented principles: Encapsulation and modularity.

The difference between a component and a subsystem is the time they are used in the software construction process. A subsystem is a design concept; it is an abstraction used in the design process to present part of a complex system. A component is a physical entity; it is an implementation tool that represents part of a bigger real system. Components are implementation realizations of subsystems. As an example, during the phases of analysis and design of a farm management system, all the functionalities related to accounting will be grouped in the subsystem *Accounting*. The collaboration with other subsystems will be defined as a set of interfaces for the subsystem. Then, functionalities of the subsystem will be translated into code and therefore the *Accounting* component is obtained ready to be deployed.

10. NOTES

A note is a symbol for rendering comments or constraints attached to an element or a group of elements [BRJ99]. Notes are used to clarify the context in which something happens. The UML symbol for a note is shown in Figure 3-14.

> text clarifying the
> context in which
> something happens.

Figure 3-14. UML symbol for a note.

Figure 3-15 shows an example of a note attached to a *Weather* class explaining the context in which the weather data are obtained. The note and the class are linked using an *anchor note*.

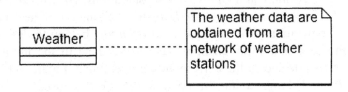

Figure 3-15. Note attached to a class using an anchor note.

11. STEREOTYPES

A stereotype is rendered as a name enclosed by guillemots and placed above the name of another element [BRJ99]. Figure 3-16 shows an example of a stereotype.

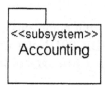

Figure 3-16. Example of stereotype.

The stereotype in Figure 3-16 is the word <<subsystem>> that gives to the package *Accounting* a special meaning or classification. Without the stereotype, the package *Accounting* is a general package. The stereotype allows the designer to create a new modeling element. Therefore, as shown in Figure 3-16, the stereotype converts a general package into a subsystem, thus creating a new building block.

Stereotypes can be used to group operations of a class into different categories, helping users to understand the context in which an operation is used. Figure 3-17 shows that operations of class *Plant* are grouped in two categories: *Initialization* and *query*. Thus, operations *setBaseTemperature*, *setFractionToCanopy,* and *setPlantDensity* belong to the category *initialization*. Operations *isMature* and *isPostPlanting* belong to the category *query*. In the case that some changes have to be done to the initialization process, it is easy to locate the corresponding operations. Figure 3-17 shows that class *Plant* belongs to the group of *entity* classes, used to represent concepts of the problem domain. (We will see more about entity classes in Part two of the book.)

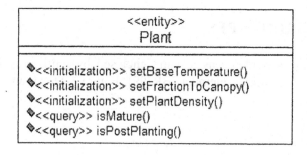

*Figure 3-17.*Using stereotypes to classify operations.

Stereotypes can be implemented in Java as comments. Figure 3-18 shows an example of using comments to translate the UML stereotypes in Java code. Figure 3-18 shows only part of the Java implementation of class *Plant*, the part related to the stereotype definition.

```
1   public class Plant {
2
3       private double baseTemperature;
4       private double fractionCanopy;
5
6       //initilaization methods
7       public void setBaseTemperature(double baseTemperature) {
8           this.baseTemperature = baseTemperature;
9       }
10      public void setFractionCanopy(double fractionCanopy) {
11          this.fractionCanopy = fractionCanopy;
12      }
13
14      //query methods
15      public boolean isMature() {
16          return getPhenologicalPhase().eqals("maturity");
17  }
18  public boolean isPostPlanting() {
19          return weather.getDayOfYear()> getPlantingDate();
20  }
21  }
```

Figure 3-18. Java comments represent UML stereotypes.

Line 6 is a comment in Java that shows the stereotype defined in Figure 3-17. Lines 7 through 12 define Java methods that belong to the category *initializations*. Lines 15 through 20 define methods belonging to the category *query*. In Figure 3-17, each operation definition includes its stereotype, whereas Figure 3-18 defines a stereotype for the entire group of methods that belong to this stereotype. In Java, methods that belong to a stereotype follow the stereotype definition.

Chapter 4

RELATIONSHIPS

As previously mentioned in this book, the main feature that distinguishes object-oriented from other programming paradigms is the fact that functionality is carried out by dialog between objects. Objects are provided with behavior that defines the role of each object. Communication between objects is realized through messages they send to each other. In UML, the ways in which things can connect to each other, either logically or physically, are modeled as relationships [BRJ99].

There are three types of relationships in UML: Associations, generalizations, and dependencies. Relationships are graphically represented with lines; each type of line represents a particular type of relationship.

1.   ASSOCIATIONS

An association is a structural relationship that specifies that objects of one thing are connected to objects of another [BRJ99]. An association shows that these two classes are connected to each other and navigation from one object to the other should be possible. This navigation is made possible through associations. Figure 4-1 shows an example of association between two classes, *Plant* and *Soil*.

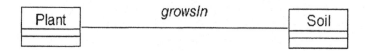

Figure 4-1. Example of association between classes Plant and Soil.

An association has a name to describe the meaning of the relationship. Names should be meaningful to present unambiguously the kind of relationship between objects. The association *growsIn* represents the fact that a plant grows in soil. The association defines quite well the role of each of the classes participating in the relationship; plant processes use soil data and soil processes use plant data. Associations enable data transfer and resource sharing among objects. In this case the association is bidirectional; data in class *Soil* can be accessed from class *Plant* and data in class *Plant* can be accessed from class *Soil*.

Accessibility between classes related with an association, in most of the programming languages, is translated with a reference of the type of the other class involved in the association. For example, access to class *Soil* from class *Plant* is possible by defining in class *Plant* an attribute of type *Soil*, which will reference the corresponding *Soil* object. In the same way, class *Soil* will have an attribute of type *Plant* pointing to the corresponding *Plant* object. In cases when both classes in a relationship point to each other, the association is bidirectional.

Although associations may have a name, some designers do not use the names, especially when the relationship between objects becomes obvious. It is recommended that all the information be provided that helps a potential user understand the presented problem.

Classes participating in an association play different roles; therefore, it is a good designing practice to explicitly define the role for each of them. Figure 4-2 shows an example of an association between classes *Plant* and *Soil* with roles defined for both classes.

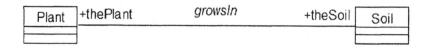

Figure 4-2. Example of an association with roles defined for both classes.

In some cases, such as in the one presented in Figure 4-2 above, it may be redundant to provide both the name of the association and the respective roles of each class. Only the name of the association may be enough to describe the nature of the relationship. The only additional information one may obtain from the names of the roles in Figure 4-2 is that the role names will be used as attribute names in respective classes. Thus, in class *Plant*, an attribute named *theSoil* will be declared to reference an object of type *Soil* (Figure 4-3). The same for the class *Soil*, an attribute of type *Plant* will be declared to reference a *Plant* object (Figure 4-4).

```
1    package Plant;
2       public class Plant {
3           Soil theSoil;
4           public Plant() {}
5       }
```

Figure 4-3. Attribute soil allows access to object Soil.

```
1    package Soil;
2       public class Soil {
3       Plant thePlant;
4       public Soil() {}
5    }
```

Figure 4-4. Attribute thePlant allows access to object Plant.

In some scenarios, only one class should have access to the data from the other class of the association. In this case, the association is unidirectional. Figure 4-5 shows an example of a unidirectional association.

Figure 4-5. Example of unidirectional association between Plant and Weather.

In the example shown in Figure 4-5, only class *Plant* can access data and behavior defined in class *Weather*. The role of class *Weather* is to provide data for class *Plant*. An attribute of type *Weather* and named *dataProvider* will be defined in class *Plant* to allow access to data and behavior in class *Weather*.

The same class can play the same role or a different one in associations with other classes. Figure 4-6 shows an example where class *Plant* plays different roles in different associations with different classes.

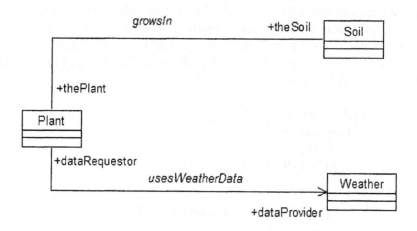

Figure 4-6. Example of class Plant playing different roles in different associations.

In Figure 4-6, class *Plant* plays the role of plant data provider in association with class *Soil* and the role of data requestor in association with class *Weather*. In Figure 4-7, class *Weather* plays the same role, the one of data provider in both associations with classes *Soil* and *Weather*.

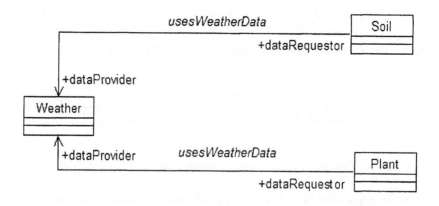

Figure 4-7. Class Weather plays the same role in two different associations.

While modeling an association, it is important to show how many objects on both sides of the association can be linked together. This process is called association's multiplicity. Figure 4-8 shows an example of multiplicity of an association.

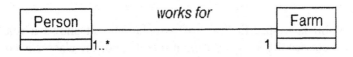

Figure 4-8. Example of multiplicity of an association.

In Figure 4-8, the association says that one or more persons work for one farm (i.e., one instance of class *Farm* can be linked to one or more instances of class *Person*).

An association can be reflexive, meaning that the start and the end of the line representing the association point to the same class. A reflexive association means that an object can be linked to other objects of the same class. Figure 4-9 shows an example of a reflexive association with multiplicity zero or one.

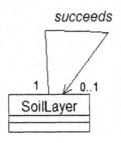

Figure 4-9. Example of a reflexive association with multiplicity zero or one.

In a soil profile, soil layers are in sequence: The first layer is on top, and other layers stay under the top layer, one under the other. The last layer of the profile is located at the bottom. The fact that soil layers are one under the other can be modeled using a reflexive association. In Figure 4-9, the association *succeeds* shows that an object of class *SoilLayer* can succeed zero or one other object of the same class. The top layer succeeds zero layer and all other layers under the top one succeed exactly one layer. A multiplicity of zero shows that the association is optional.

2.   AGGREGATION

Aggregation is a specific kind of association that shows a relationship between two classes that play a different role. One of the classes is considered the *whole* and the other one is considered the *part*. An aggregation expresses a whole-part relationship. Figure 4-10 shows the UML representation of an aggregation.

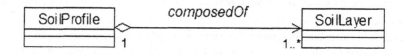

Figure 4-10. The UML representation of an aggregation.

The association presented in Figure 4-10 shows the relationship between a soil profile and its soil layers. The association is a one-to-many; meaning that a soil profile is composed of one or more soil layers. The class *SoilProfile* represents the whole and class *SoilLayer* represents the part. The open diamond distinguishes the whole class. It is important to note that in an aggregation, the whole does not own the part. Therefore, when the class representing the whole is destroyed, this does not affect the class representing the part. Furthermore, an object representing the part may be used in other aggregations. An aggregation is an association that represents a whole-part relationship in a conceptual way.

As aggregation expresses the whole-part relationship in a very loose manner, many modelers do not see it as useful modeling concept. A relationship expressed by an aggregation can be modeled using a simple association. In this case, the name of the association could express the idea of whole-part relationship. Choosing between an aggregation and a simple association for conceptually expressing the whole-part relationship is a question of taste or modeling habits.

A one-to-many association can implemented in Java as an array. Line 4 in Figure 4-11 shows that the attribute *soilLayer* is an array that will contain objects of type *SoilLayer*.

```
1   package Soil;
2
3   public class SoilProfile {
4       public SoilLayer soilLayer[];
5       public SoilProfile() {}
6   }
```

Figure 4-11. Java implementation of a one-to-many association.

There is a large class of water-balance and irrigation-scheduling models that requires modeling the relationship between the soil profile and its soil layers [PSH04]. Some models do not partition soil into layers; they simply consider soil profile as a single layer that extends to the bottom of the root zone [GSR00]. In these cases, the model will always have one layer.

Other water-balance and irrigation-scheduling models consider a soil profile as composed of many soil layers [Rit98]. Therefore, in these models, one soil profile will be associated to many soil layers. The association in Figure 4-10 takes into consideration both cases.

3.   COMPOSITION

A composition is a stronger form of aggregation, with strong ownership and coincident lifetime as part of the whole [BRJ99]. In a composition association, the whole is responsible for the creation and destruction of its parts. Once a part is created, it belongs to the whole and when the whole is destroyed, the part is destroyed too. In a composition, a part may belong to only one whole at a time.

As an example, let us consider constructing a UML diagram that represents a plant and its relationships with its root, stem, and leaves systems. The relationship between plant and its systems can be presented as a composition. Plant will play the role of the whole and its systems will be its parts. Plant owns its root, stem, and leaves systems. A stem system can only belong to one plant at a time. In most cases if the plant dies, so do its root, stem, and leaves systems. The UML presentation of a composition is shown in Figure 4-12.

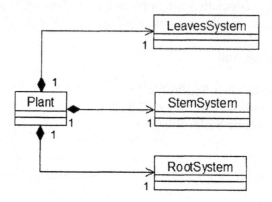

Figure 4-12. Plant is considered as a composition of its root, stem, and leaf systems.

4.   DEPENDENCY

A dependency relationship states that a change in specification of one thing may affect another thing that uses it, but not necessarily the reverse [BRJ99]. Figure 4-13 shows an example of dependency between packages *Client* and *Supplier*. *Client* depends on *Supplier*. If the amount of goods that a *Supplier* is supposed to provide changes, this change may affect the *Client* as the *Client* would have to adjust its behavior to accommodate the change.

Figure 4-13. Dependency relationship between packages.

A dependency relationship can exist between classes, packages, and components. The following Figure 4-14 shows an example of dependency relationship between components *Client* and *Supplier*.

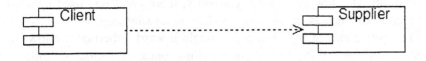

Figure 4-14. Dependency relationship between components.

In some cases, one modeling element can be dependent on more than one other modeling element. Figure 4-15 shows that the amount of water in soil depends on two factors: The weather and the irrigation applied.

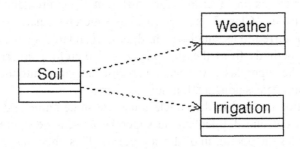

Figure 4-15. One class depends on two other classes.

A dependency relationship denotes a relationship where the client does not have semantic knowledge of the supplier.

5. GENERALIZATION

A generalization is a relationship between a general thing (called the superclass or parent) and a more specific kind of that thing (called subclass or child) [BRJ99]. Generalization is the relationship that represents the mechanism of inheritance in object-oriented languages.

Inheritance is often considered as one of the most fundamental features of the object-oriented paradigm. It is certainly the feature that distinguishes object-oriented from the traditional programming. Inheritance was introduced to the world of programming in the late 60s as the main feature of the programming language SIMULA [DMN68]. SIMULA's inheritance mechanism was originally known as *concatenation* and the term *inheritance* was introduced a few years later. Currently, there are a few used synonyms for inheritance, such as *subclassing, derivation,* or *subtyping.*

The central idea of inheritance is straightforward. Inheritance allows new object definitions to be based upon exiting ones. A formal definition of inheritance is given by [BC90]:

R = P + dR

P are the properties inherited from an existing class and **dR** are the new properties defined in the new class **R**. **dR** represents the incrementally added properties in class **R** that make class **R** different from class **P**. The symbol + is some operator that combines the exiting properties with the newly added ones. Inheritance is a facility for differential, incremental program development [Tai96]. Class **P** is referred to as superclass, parent, or ancestor and class **R** is referred to as subclass, child, or descendant. A subclass inherits attributes and methods from the superclass, and therefore inherits data and behavior from the superclass. As such, a subclass can substitute the superclass anywhere the superclass appears, but not vice versa.

If an operation defined in the subclass has the same name and parameters or signature as the one defined in the superclass, then the operation of the subclass overrides the operation of the superclass. This phenomenon is known as polymorphism.

A subclass can even cancel an operation defined in the superclass. This can be achieved by simply redefining the same operation and not providing any logic for it. Generally a subclass may introduce new properties in addition to the ones defined in the superclass that extend, modify, or defeat them [Tai96]. The UML notation for the generalization is shown in Figure 4-16.

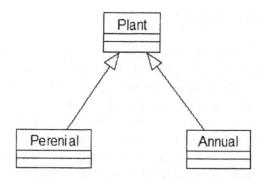

Figure 4-16. UML notation for the relationship of generalization.

In Figure 4-16, perennial plants are a special case of plants and annual plants are a special case of plants, but perennials are different from annual plants. A perennial is a kind of plant, a specific kind of plant. A plant considered randomly may not necessarily be a perennial. The generalization relationship expresses a certain hierarchy of objects; moving up in the hierarchy objects become more general and moving down in the hierarchy objects become more special. Any object at a lower level can replace an object residing higher in the hierarchy.

Inheritance is the mechanism by which more-specific elements incorporate the structure and behavior of more-general elements [BRJ99]. Both terms, generalization and inheritance, are generally interchangeable but there is a clear distinguishing between them. Generalization is the name of the relationship, whereas inheritance is the mechanism that the generalization relationship represents.

In Figure 4-17, a hierarchy of different classes related by inheritance is shown [AR97]. The definition of class *ShootOrgan* includes a number of attributes that will be inherited by *Stem*, *MainStem*, and *BranchStem*. For example, *age* is an attribute defined at the *ShootOrgan* class and classes *Stem*, *MainStem*, and *BranchStem* will have an attribute with the same name although it is not shown in their attribute compartment. A subclass shows only the attributes defined at subclass level. Therefore, objects created by *MainStem* class will have attributes defined by *ShootOrgan*, *Stem*, and *MainStem* classes. The list of attributes of class *BranchStem* will contain *on_stem_number*, *location_onStem*, *length*, and *number_leaves_on_stem* defined in the abovementioned classes.

The author, [AR97], has chosen not to provide any behavior for classes *ShootOrgan* and *Stem*. Therefore, objects created by class *MainStem* will not inherit any behavior from the superclasses *ShootOrgan* and *Stem*.

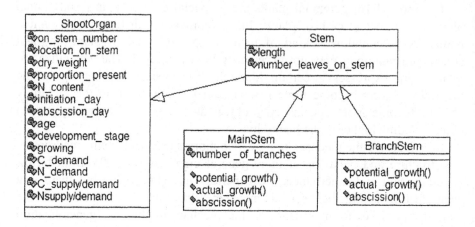

Figure 4-17. Example of a hierarchy of classes related by inheritance.

Generalization is transitive; class *Stem* inherits from class *ShootOrgan* and class *MainStem* inherits from *Stem*. Therefore, class *MainStem* inherits from class *ShootOrgan*.

Another example of inheritance is taken from Lemmon [LCh97]. Classes shown in Figure 4-18 are all organs, as they inherit directly or indirectly from class *Organ*. Class Leaf directly inherits data and behavior from class Organ. Classes *SympodialLeaf*, *PreFruitingLeaf*, and *FruitingNodeLeaf* inherit data and behavior from class *Leaf* and at the same time, they inherit from class *Organ*. A *PreFruitingLeaf* object is a specialized kind of *Leaf* and a specialized kind of *Organ*.

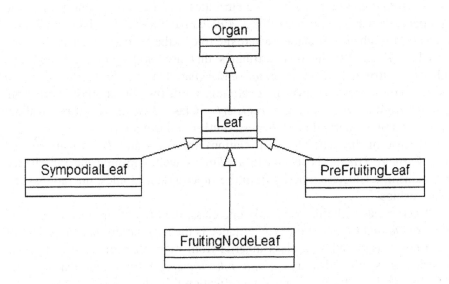

Figure 4-18. Hierarchy of classes in a cotton simulation model.

As previously mentioned, subclasses inherit from superclasses data and behavior. Figure 4-19 shows an example of what is inherited through a generalization relationship.

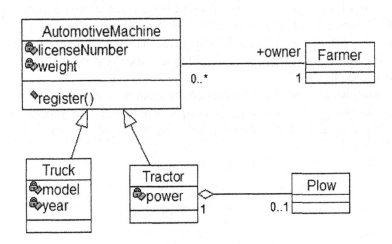

Figure 4-19. A subclass inherits from the superclass attributes, operations, and relationships.

An *AutomotiveMachine* has two attributes: *licenseNumber* and *weight* and an operation named *register()*. Class *AutomotiveMachine* is related with class *Farmer* through an association where the role of the farmer is *owner*.

Class *Truck* has its own attributes that are *model* and *year* and two inherited attributes from *AutomotiveMachine* that are *licenseNumber* and *weight*. Truck does not have any operations on its own but it does inherit from *AutomotiveMachine* operation *register()*. Class *Tractor* has the attribute *power* and two inherited ones, *licenseNumber* and *weight*.

Because of the generalization relationship, *Truck* and *Tractor* are related to a *Farmer*. *Tractor* is related to a *Plow* as well. Through the inheritance mechanism, a subclass inherits from the superclass attributes, operations, and relationships.

When a class inherits from only one class, then the inheritance is referred to as *single inheritance*. Single inheritance is the most common mechanism of inheritance used. When a class inherits from more than one class, then the inheritance is referred to as *multiple inheritance*. Multiple inheritance offers more possibilities for incremental modification than the single inheritance, but its use is not easy. Almost unanimously, researchers agree that the use of multiple inheritance should be done with care as its use introduces technical and conceptual problems. Some authors state that despite the problems the use of multiple inheritance raises, any modern object-oriented language should provide support for it [Taiv96]. Other researchers do not agree; they state that multiple inheritance is dangerous and should not be used. A strong support up for the argument against the use of multiple inheritance can be found in the lack of its implementation in two of the modern object-oriented languages; Java and C# provide a single inheritance mechanism while C++ provides support for multiple inheritance. Multiple inheritance is good, but there is no good way to do it [Coo87].

All examples presented in this section are examples of single inheritance. Figure 4-20 shows an example of multiple inheritance. As shown in this figure, class *Segment* inherits data and behavior from both classes *Article* and *Organ* [DP01].

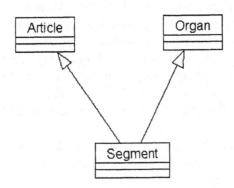

Figure 4-20. Class Segment inherits from both classes Article and Organ.

6.   ABSTRACT CLASSES

During the discussion of the mechanism of inheritance, it was mentioned that general classes stand at the top of the hierarchy. As we go up in the class hierarchy, classes become more and more general. In this context, some of the classes standing at the top of the hierarchy can play a specific role which is the role of defining some behavior that would be common to all other classes lower in the hierarchy. Such a class does not fully represent any object; it only represents a template for creating other objects that will have in common the behavior defined in this class. As such, these classes are abstract; there are no direct instances created from them. Abstract classes represent incomplete abstractions that can be useful for specifying contracts upon which concrete implementations will be based. These abstractions have a communicative role that allows designers to agree on interface specifications before starting concrete implementations. An abstract class is written with the expectation that its subclasses will add to its structure and behavior [BRJ99]. Part of designing an abstract class is specifying how this behavior will be used or refined by other classes lower in the hierarchy.

One of the classical examples of inheritance is taken from the domain of animals. At the top of the hierarchy a class named *Mammals* would be defined. According to The Pocket Oxford Dictionary, a mammal is "a warm-blooded vertebrate of the class secreting milk to feed its young." The fact of feeding the young with milk is the most common thing that all animals have. It makes sense to create a class named *Mammals* that will capture this common behavior of all animals. However, there are no direct instances of the class *Mammals*, as mammal is a general concept representing a group of animals with common characteristics. Therefore, *Mammals* is an abstract class that has no instances and provides the most common data and behavior all animals have. Dog is a concrete animal that shares the characteristics described in *Mammals*. A Dog is a mammal and to represent a dog, a concrete class needs to be created inheriting from the abstract class *Mammals*. All other animal classes would inherit from the *Mammals* class. Each class representing a particular animal will extend the data and behavior defined at the abstract class level.

The UML presentation of an abstract class is the same as the one used for other classes, but the name of the class is in italicized font. Figure 4-21 shows an example of an abstract class.

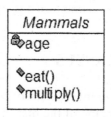

Figure 4-21. Example of an abstract class. The name of the class is written in italics.

As previously mentioned, abstract classes do not have any instances. The attributes and the operations defined in an abstract class are inherited by all classes that are a specialization of the abstract class, meaning all classes that subclass the abstract class. Figure 4-17 shows an example of abstract classes created in a hierarchy of classes to address issues of developing a generic object-oriented plant simulator [AR97]. In this example, class *Organ* is an abstract class and its definition contains only attributes, no operations. In the same way, class *Stem* inherits from *Organ* but does not define any behavior either. Classes *MainStem* and *BranchStem* define additional attributes and the same behavior, such as *potential_growth()*, *actual_growth()*, and *abscission()*.

Let us look more carefully at this example of hierarchy as there are some important points to be made. When designing a class hierarchy, attention should be paid to the fact that every class of the hierarchy should have a well-defined purpose and role that will make the class have a specific place in the problem domain. Class's behavior is used in the dialog with other classes to achieve functionality. Classes should not be created just for being a place-holder of some values; they should play a specific and well-defined role in the domain in study. A class is a behavioral template for its instances [Zdo99].

A class without behavior generally causes difficulty of use and justification of need for such a class. Furthermore, inheriting from a class without behavior makes the problem even more complicated. As class *Stem* subclasses *Organ*, then *Stem* is a kind of *Organ*. Or class *Organ* does not have any defined behavior so stating that *Stem* is a kind of *Organ* does not provide additional information about class *Stem*. What is an *Organ*? What is the role of class *Organ*? How does class *Organ* dialog with other classes? Assigning a meaningful name to a class is not enough. A class should be the product of abstraction used to depict potential players in the domain under study. Selecting a meaningful name is important, as it enables other people to understand more easily the purpose and the role of the class, but a name is not sufficient. The name of the class by itself does not provide any behavior; it is only a label that distinguishes a class among others in the problem domain. Each class should provide some specific functionality that is not already defined in other classes.

What kind of behavior can class *Organ* provide? Class *Organ* should be provided with the most common behavior organs of a plant have. Organs grow or die, they actively interact with the surrounding environment. The role that organs play in collaborating with other organs or parts of the environment should be defined in class *Organ* in an abstract way. This behavior can be detailed or redefined by other classes inheriting from class *Organ*.

Furthermore, Figure 4-17 shows that both classes, *MainStem* and *BranchStem*, at the end of the hierarchy provide the same behavior (i.e., the ability to calculate potential and actual growth and abscission). Both classes inherit from class *Stem*, and *Stem* does not provide any behavior; it only adds two more attributes (length and number_leaves_on_stem) to the list of attributes defined at *Organ* class. Again, class *MainStem* inheriting from *Stem* is a kind of *Stem*. *Stem* does not provide any behavior nor does it inherit from *Organ*. Saying that a *MainStem* is a kind of *Stem* does not provide any information about what *MainStem* is or how it does behave.

In Figure 4-17, it is shown that *MainStem* and *BranchStem* are provided with behavior (operations *actual_growth* and *potential_growth*) that allows them to grow. A *Stem* is subject to potential and actual growth too; therefore, this behavior should be moved up to *Stem* level and each of the subclasses (*MainStem* and *BranchStem*) can provide polymorphic behavior for potential and actual growth, and abscission.

Abstract classes have a twofold role. The first role is a conceptual one; abstract classes can serve as modeling tools or specifications with the aim of identifying abstractions of the problem domain that will later be refined by concrete classes lower in the hierarchy. The second role is more of a utilitarian nature; abstract classes can serve as templates for improving reusability. The behavior defined at abstract classes will be implemented by concrete classes that subclass it [Tai96].

Abstract classes are a useful design technique that promotes code reuse. Depicting abstract classes in a problem domain is an iterative process that uses abstraction to find common functionalities in concrete classes and move it to a higher level in the hierarchy. The advantage of using abstract classes is that behavior common to many classes can be defined in only one place to be reused, modified, or improved later.

7.   ABSTRACT CLASSES VERSUS INTERFACES

The concepts of abstract classes and interfaces are somewhat similar and one might be confused in deciding whether to use an interface or an abstract class. An abstract class defines a default behavior for some or all the operations that will be inherited by all the subclasses. The reuse of the behavior defined in an abstract class is realized through inheritance.

An interface does not define any default behavior at all. Interfaces only define specifications that will be implemented by classes realizing the interface. An interface may be implemented by many classes, and a class can implement many interfaces.

8.   REALIZATION

A realization is a semantic relationship between classifiers in which one classifier specifies a contract that another classifier guarantees to carry out [BRJ99]. The most common use of the realization is between interfaces and the classifiers that agree to implement the interfaces. Figure 4-22 shows examples of realizations.

Figure 4-22. Examples of realizations.

As shown in Figure 4-22, a realization is an agreement between a class and an interface. The interface defines the functionalities that the class should provide the implementation. The realization is presented by the line that connects the interface and the class. In the same way, a realization connects an interface that defines functionalities and the subsystem that would provide the implementation. Figure 4-23 shows another type of notation for realization.

Figure 4-23. Another notation for realization.

Chapter 5

USE CASES AND ACTORS

Usually software systems are developed to be used by humans or other hardware devices. There is a close interaction between users (humans or machines) and the system. Users send a message to the system that provokes the system to execute some operations in order to return some valuable response. Therefore, determining what a software system should provide to users means understanding what the users want from the system. The process of capturing requirements for a system developed using object-oriented approach is referred to as developing the use case model.

The use case approach was introduced by the well-known work of [JCJ94], often referred to in the object-oriented community as the father of the Object-Oriented Software Engineering (OOSE). Very soon, use cases were embraced by the totality of the methodologist worldwide.

Use cases are a simple and yet powerful way to express the functional requirements of a system. Use cases describe how users can use the system and what the system can do for users. Therefore, use cases are an important tool to build a consensus between the system's stakeholders and the system's developers. If stakeholders cannot agree on what the system should provide, chances that the project can be successful are very slim. Use cases have improved the communication between stakeholders and the development team and have made the process of gathering system requirements easier and more formal. Use cases provide a visual representation of the conceptual model of the system. More details about use case modeling can be found in [JCJ94], [BS03], and [BRJ99].

The use case model contains *actors* that represent the future users of the system and *use cases* that represent what the users can do with the system.

1.   ACTORS

An actor represents a coherent set of roles that users of use cases play when interacting with these use cases [BRJ99]. Actors represent the role of the future users of the system. Actors model the user's perspective of the system. Actors are located outside the system; therefore, in order to depict actors, it is important to define the boundaries between actors and the system.

The UML symbol for an actor is shown in Figure 5-1. An actor has a name that distinguishes actors among them. It is a good modeling practice to name actors by the role they play, not by their names. The name of a person may change but this will not affect the role this person plays in the system.

Farmer

Figure 5-1. The UML symbol for an Actor.

There are three primary types of actors: Users of the system, other systems interacting with our system, and time [BB02].

The first type of actor is a person or a user who will use the system. These are the most common type of actors. As an example, in a crop simulation scenario, a farmer will ask the system to run a simulation and therefore, the farmer is an actor.

The second type of actor is another system interacting with our system. For example, the crop simulation system obtains the weather data directly from a weather station on-line. In this case, the weather station is outside our system and it is not our intention to modify its behavior; therefore, the weather station is an actor.

The third type of actor is time. Time becomes an actor when after a certain period of time, a series of events to be handled by the system is triggered. As an example, an advisory system can be designed to function based on weather conditions. When weather conditions (temperature and humidity) favor development of certain diseases or fungus, the system will provide advice for starting spraying with appropriate pesticides.

The difference between actor and user of the system is rather subtle; a user is someone that uses the system, whereas an actor is a role that a user can play. A user can play several roles and therefore a user can be modeled as different actors.

Actors can be linked to each other using the generalization relationship. Figure 5-2 shows an example of generalization between actors. A Commercial Customer is a special case of a Customer, i.e., Commercial Customer inherits from Customer. Although actors are outside of the system and not the subject of our study, it is useful to know how they are structured and related, as it helps to understand how they communicate with the system.

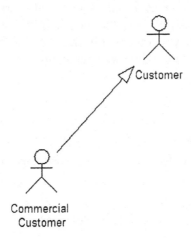

Figure 5-2. Actors related using a generalization relationship.

2. USE CASES

Originally, Jacobson [JBR98] defined a use case as "a behaviorally related sequence of transactions in a dialog with the system." A more recent definition of the use case is given by [BRJ99] as "a description of a set of actions, including variants that a system performs to yield an observable result of a value to an actor." The basic idea behind a use case is to represent a sequence of interactions between the system and its users located outside the system. In other words, a use case shows how an actor uses a system to achieve a certain goal and what the system should do for the actor to achieve that goal. It describes how the actor and the system collaborate to deliver a result of value to the actors [BS03].

Use cases are widely accepted to be the best practice for capturing system requirements [Kru98]. Functional requirements capture the intended behavior of the system. The use case model expresses the functionalities the system is supposed to provide to its users. Use cases only specify how the system should behave; they do not specify how the behavior should be implemented. Therefore, use cases are considered to be an excellent way of communicating with customers and users of the system.

The UML symbol for a use case is shown in Figure 5-3. The use case *Simulate* only shows that users should be able to send the message *simulate* to the system and the system will execute all the necessary operations. For the moment, how the simulation will be achieved is not important. The same way, the use case *Get Weather Data* only shows that users may ask the system to carry out this functionality; how the data will be obtained is not relevant at this point. The data may be obtained from reading a data file or a database, or obtained directly from an on-line weather station.

Simulate   Get Weather Data

Figure 5-3. Example of use cases.

Use cases have names that distinguish them from each other. Usually, use case names are of form <Verb><Noun>, such as *Get Weather Data, Simulate*, that shows that users are making a request to the system and the system should provide back some results.

An actor and a use case are related through an association as shown in Figure 5-4. The actor (*Farmer*) initiates the use case by sending a message to the use case. Use cases are always started by actors. The communication shown in Figure 5-4 is unidirectional, as it goes from the actor to the use case. The sense of the communication is clear; it goes from the actor to the use case.

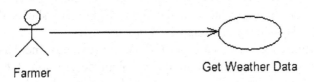

Figure 5-4. Farmer asks the system for weather data.

In Figure 5-5, the association linking the actor and the use case is bidirectional; it is not clear whether the actor initiates the use case or the use case communicates with the actor. It is important to clearly describe the type of communication between actors and use cases, as it helps to better understand how the system works.

Figure 5-5. Example of a bidirectional use case.

2.1 Extend relationship

An extend relationship between use cases means that the base use case implicitly incorporates the behavior of another use case at a location specified indirectly by the extend use case [BRJ99]. The base use case must be defined to completely stand by itself. Its description should be independent of the use case that extends it. The extend use case will be executed only when some particular circumstances will be satisfied in the base use case. Extended use cases can be successfully used to add additional functionalities to base use cases without questioning their integrity.

Let us consider as an example the process of approving extension documents (http://ers.ifas.ufl.edu) at the Institute of Food and Agricultural Sciences (IFAS), at the University of Florida. Each document needs to be peer-reviewed by at the least two reviewers, before it goes for approval to the department chair. According to this practice, most of the time the reviewers selected to review the document are sufficient. The department chair can add additional reviewers in the case that the document deals with issues that none of the reviewers is a specialist in the field. Figure 5-6 shows the use case model for the department chair approval process.

In this figure, the use case *Add Reviewer* extends the base use case *Approve Document*. According to the problem description, the department chair can add an additional reviewer to a document when he judges that a more specialized reviewer should review the document. This means that normally, the department chair considers that reviewers assigned to the document are sufficient. Thus, the description of the base use case *Approve Document* is independent of the use case *Add Reviewer*. The base use case can be executed without involving the extend use case. The functionality provided by the extend use case *Add Reviewer* is needed only under certain conditions, when the department chair finds it necessary.

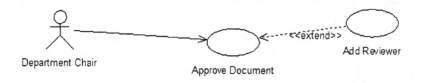

Figure 5-6. Use case Add Reviewer extends base use case Approve Document.

2.2   Include relationship

An include relationship between use cases means that the base use case explicitly incorporates the behavior of another use case at a location specified in the base [BRJ99]. An include relationship represents a set of operations that are repeated in several use cases and are grouped in one place for ease of use and maintenance. An included use case never stands by itself; it is always instantiated as part of a larger use case.

Let us consider again IFAS's extension documents approval system. One of the requirements was to develop an event-based system. Every time an event occurs, the next person in the approval process should be automatically notified. When an author submits a document for approval, the department chair gets immediate notification. When the department chair approves a document, the program leader gets immediate notification.

In Figure 5-7, the base use case *Approve Document* includes the use case *Notify*. The *Notify* use case represents a group of operations needed to send a notification message to anyone interested to know that some event has happened. Therefore, this set of operations is repeated in several places. It is convenient to group this functionality in one place, and any other use case that needs it can use it by simply including it in its definition.

Figure 5-7. Base use case Approve Document includes use case Notify.

Note that both relationships *extend* and *include* use a dependency relationships. In the case of the *extend* use case, *AddReviewer* depends on *Approve Document*, as it is at the discretion of the department chair whether to add an additional reviewer or not. In the case of *include* use case, *Approve Document* depends on *Notify* to go to the next level of approval. A stereotype (i.e., *extend* or *include*) is used to show the type of the use case.

Chapter 6

UML DIAGRAMS

UML provides five kinds of diagrams for modeling the dynamic aspects of systems. These diagrams are: Use case diagrams, sequence diagrams, collaboration diagrams, activity diagrams, and statechart diagrams. Use case diagrams are central to model the behavior of a system.

1. THE USE CASE DIAGRAM

The set of all use cases in a problem domain is referred to as the *use case model* and the diagram representing it is referred to as the *use case diagram*. A use case model shows the set of functionalities a system should provide. By examining a use case model, we can say whether all the user requirements are satisfied or not. A use case model is important, as it presents a general view of the system without being overwhelmed by implementation details.

Let us consider IFAS's extension document approval system and build the use case model. The following is a brief description of the functionalities the system should provide.

The system should allow users (authors, editors, reviewers, department chairs, and program leaders) to submit, edit, review, approve, and check the status of a document any time. First, the authors should submit the document and then the editor edits it. At the least, two reviewers, assigned to the document, will be notified for reviewing the document. After the reviewer's approval, the department chair is notified to approve the document. If the department chair judges that another and more specialized reviewer should review the document, then a new reviewer can be added. The newly added reviewer will be notified by mail that a document is waiting for approval.

After the additional reviewer approves the document, the department chair is notified by mail that the document is waiting for departmental approval. When the department chair approves the document, the program leader is notified by mail. When the program leader approves the document, then the document is saved in the database and indexed. It becomes public and available for search purposes. The use of the system should be password protected.

The use case model should express all the functionalities required by the system. By examining the use case model one should be able to judge whether all users requirements are correctly captured and whether all user's roles are included in the system. Figure 6-1 shows a simplified use case model for a tracking system for extension documents (http://ers.ifas.ufl.edu).

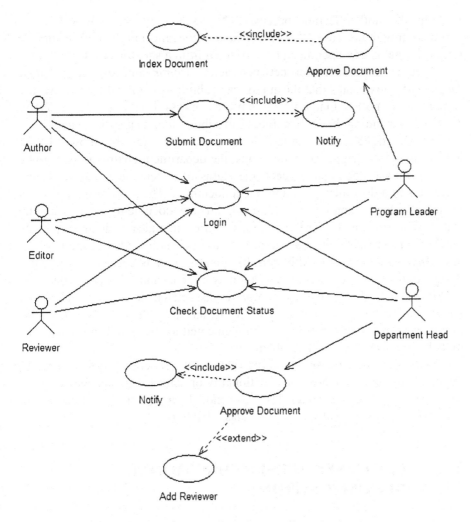

Figure 6-1. The use case model for a document tracking system.

Let's closely examine each of the use cases presented in the use case model and evaluate whether all the users requirements are correctly captured.

The use of the system is password protected; therefore, all users need a password to log into the system. The authentication process not only validates the user, but also its role. When a user logs into the system with a specific role, the user has access to the functionalities that the role is entitled to have. The operations required to verify whether a user is a legitimate one are presented by the use case *Login*. There is an association between each of the actors and the use case *Login*; this means that every user of the system goes through the authentication process. The same way, all actors have an association with use case *Check Document Status*, meaning that all actors can check the status of a document, each of them from a specific point of view.

There is an association between actor *Author* and use case *Submit Document*. This means that the author can submit a document for review. It is important to notice that only *Author* has access to the use case *Submit Document*, meaning that only authors can submit documents for review.

The functionalities a certain actor is entitled to are shown by all associations that origin this actor. Thus, the department chair can login to the system, browse documents, check the status of a document, add additional reviewers to a document, and approve a document. The same way, a program leader can login, approve a document, browse documents, and check the status of a document. The operations needed to index a document in the database are presented as a separate use case that is included in the base use case *Approve Document*. Although this functionality is used only once, it is designed to be a separate use case, as it may be reused in other occasions.

Figure 6-1 shows that *Reviewer* is associated to *Login* and *Check Document Status* use cases only. This means that *Reviewer*'s role is not designed properly; if the system is implemented as presented in the use case model, *Reviewer*'s role is incomplete.

The presented use case model provides all the functionalities required by the users. In the case that some functionality or actor is not considered, it can easily be added to the model. Use case models are central to modeling the behavior of a system, subsystem, or a class [BRJ99].

2.   USE CASES VERSUS FUNCTIONAL DECOMPOSITION

Often, use cases are confounded with a detailed list of functions the system should provide. When this happens, use cases are defined as if they represented menu items of a system. Figure 6-2 shows an example of designing use cases as menu items. At the heart of the diagram, shown in Figure 6-2, is an actor that initiates three use cases; *Modify File, Add File*, and *Delete File*.

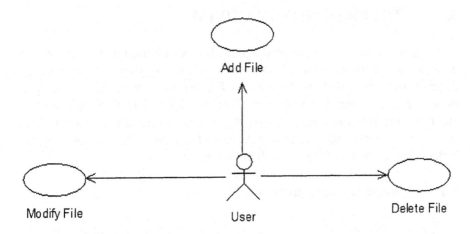

Figure 6-2. Example of bad selection of use cases.

Figure 6-2 shows things that the systems should do, but they are all related to only one thing the user wants the system to do: Administer a file system. According to the definition of the use cases, they describe what the system should do that will benefit at the least one of the actors. The use case *Delete File* may never be invoked if a file has not been added to the system. The same reasoning can be done for the use case *Modify File*. Events like modifying and deleting a file are useful to a user only when a file is already added to the system. Therefore, all three functions can be gathered in a sole use case named *Administer Files* as shown in Figure 6-3. Gathering several functions into a unique use case is a better presentation of what the system should do for the users and it focuses on the value the user will obtain from the system. Dividing the behavior of the system into small functionalities does not help to understand the conceptual model of the system.

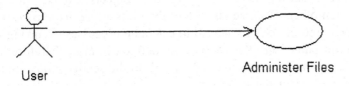

Figure 6-3. A use case represents a set of functions.

3.   INTERACTION DIAGRAMS

UML uses different types of diagrams for expressing the dynamic aspects of systems; the use case model diagram is only one of them. Another type of diagram used by UML is *Interaction Diagrams*. An interaction diagram shows an interaction, consisting of a set of objects and their relationships, including the messages that may be dispatched among them [BRJ99]. They are used to capture the dynamic behavior of a system. Interaction diagrams include sequence diagrams and collaboration diagrams.

3.1   Need for interaction

The use case diagrams describe the system, the surrounding environment, and the relationships between them. Actors are located outside the system and they start a request. The system receives the request and executes all operations needed to provide the actor with a response. As previously mentioned, the use case model presents the entire set functionalities the system should provide to its users.

The most important concepts of the problem domain are represented as classes. Classes are provided with data and behavior so they can play a well-defined role. Classes are factories for producing objects. The system is composed of objects that interact with each other to achieve functionality. Objects dialog between themselves through messages.

A message is the specification of a communication among objects that conveys information with the expectation that activity will ensue [BRJ99]. Messages are the mechanism that allows objects to interact with each other. Objects make their behavior available to others through messages. A message is a call through which an object asks another object to do something. The object receiving the message will execute it and may give the result back to the object sending the message.

Figure 6-4 shows an example of two objects exchanging messages between each other. *Plant* sends to *Soil* the message *getWaterStress()*. Water stress data stored in *Soil* are needed to calculate processes occurring in *Plant*. An operation named *getWaterStress()* is defined in *Soil*; therefore, *Soil* will execute the operation and provide the result to the sender *Plant*. In the same way, processes occurring in *Soil* need leaf area index data that are located in *Plant*. Therefore, *Soil* sends a message to *Plant* asking for leaf area index data. Plant receives the message, executes the operation named *getLeafAreaIndex*, and provides the result to the sender *Soil*. The operation *getLeafAreaIndex* is part of *Plant* behavior and operation *getWaterStress* is part of behavior of *Soil*. The return result of the operation *getWaterStress()* is of type *double*, as shown by the signature of the operation in Figure 6-4. The type of the return result is needed in the calculations occurring in *Plant*. In the same way, the type of the return result *getLeafAreaIndex* is a *double*. The sender should be aware of the type of the return result, as it might be used for further calculations occurring in the sender object.

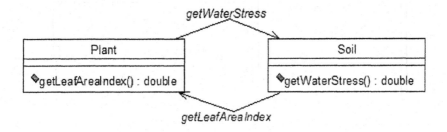

Figure 6-4. Interactions between Plant and Soil.

3.2   Sequence diagrams

A sequence diagram is an interaction diagram that emphasizes the time ordering of messages [BRJ99]. A sequence diagrams represents objects participating in the interaction in a timely manner. The time when messages are sent to objects is important and altering this order may produce unexpected results.

Figure 6-5 shows an example of a sequence diagram. Farmer plays the role of an actor as the farmer sends a request to the system to obtain some weather data.

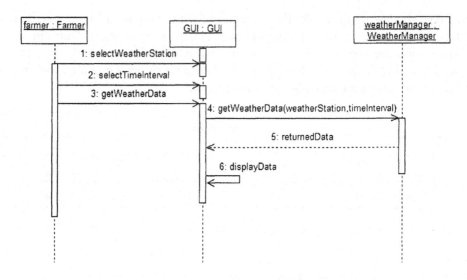

Figure 6-5. Example of a sequence diagram.

Farmer communicates with object *GUI* (Graphical User Interface). The sequence of messages sent between the objects described in Figure 6-5 is as follows.

First, the farmer needs to select a weather station from the list displayed by the GUI object. Second, the farmer needs to select a time interval for the weather data. Third, the farmer needs to press the button *GetWeatherData*. Object *GUI* sends the message *getWeatherData(weatherStation ,timeInterval)* to object *WeatherManager.* Note that this message has two parameters selected by the farmer: The weather station and the time interval for the data. *WeatherManager* will execute the message and return the data to object *GUI* that sends himself the message *displayData*. The farmer can then read the displayed data.

In this particular example, the order of the first two messages the farmer sends to GUI can be reverted; the farmer may select the time interval first and then select a weather station. It is understandable that message 4, *getWeatherData(weatherStation, timeInterval)*, cannot be executed before message 3, *getWeatherData*.

In a sequence diagram, objects are shown as vertical lines as shown in Figure 6-6. The vertical line is referred to as object's lifeline. The lifeline shows when an object is created and how long its life would be. Lifelines are used to model class behavior. Figure 6-6 shows that *Farmer* is the name of the class and *farmer* is an object of class *Farmer*.

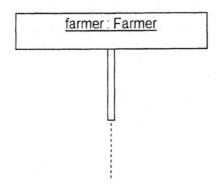

Figure 6-6. Specifications on sequence diagrams.

Objects communicate among them by sending messages. Figure 6-7 shows an example of objects sending messages to each other. A message is shown by an arrow. A message has a sender, which in the case of Figure 6-7 is object *farmer*, and a receiver, which is object *GUI*. When object *farmer* sends a message to object *GUI*, it means that *farmer* needs to use some of the behavior defined in object *GUI*.

When an object receives a message, it need some time to execute the message and send the results to the sender. The time during which an object is performing an operation is referred to as the *focus of control* [BRJ99]. In Figure 6-7, object *GUI* receives a message and it has the focus of control.

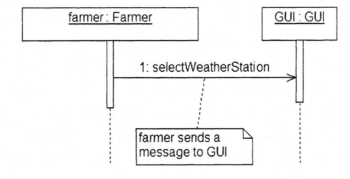

Figure 6-7. Object farmer sends a message to object GUI.

In some cases, an object can send a message to itself; these messages are referred to as reflexive messages. The sender and the receiver of the message, in this case, is the same object. Figure 6-8 shows an example of a reflexive message; object *GUI* sends to itself the message *displayResults*.

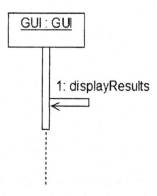

Figure 6-8. Example of a reflexive message.

3.3   Collaboration diagrams

[BRJ99] defines a collaboration diagram as "an interaction diagram that emphasizes the structural organization of the objects that send and receive messages; a diagram that shows interaction organized around instances and their links to each other." An example of a collaboration diagram is shown in Figure 6-9.

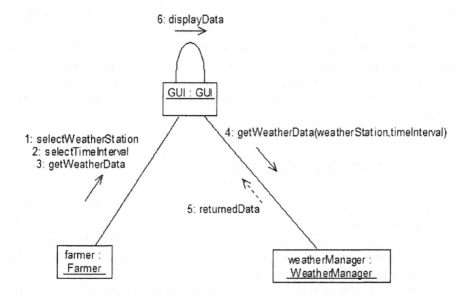

Figure 6-9. Example of a collaboration diagram.

In a collaboration diagram, all the messages that start at an object and the ones that end at an object are shown. As shown in Figure 6-9, messages *selectWeatherStation*, *selectTimeInterval*, and *getWeatherData* start from object *farmer*. Message *displayResults* starts and ends at object *GUI*.

Messages that end at an object show that the behavior of this object should be designed to provide answers for the received messages. Therefore, collaboration diagrams help design class behavior. From the collaboration diagram presented in Figure 6-9, class *GUI* should, at the least, provide behavior for selecting a weather station, selecting a time interval, and to get the weather data from the source used.

3.4   Sequence versus collaboration diagrams

Sequential and collaboration diagrams are semantically equivalent. They express the same thing: The interaction between objects. It is easy to convert one diagram to the other, as they present the same information. Some UML software, such as Rational Rose, provide automatic conversion from one diagram to the other.

Although both diagrams present the same information, they do not visualize the same information. Sequence diagrams are used when modeling a flow of control over time and when it is important to represent the messages passed between objects, as they unfold over time. Therefore, sequence diagrams are very useful to describe use case scenarios.

Collaboration diagrams are used when modeling a flow of control by organization and when it is important to emphasize in the structure of the relationships between objects and in the totality of messages an object may receive. Therefore, collaboration diagrams are used to build class and object behavior.

4.   ACTIVITY DIAGRAMS

An activity usually represents a set of actions where execution may cause a change in the state of the system or return a result. An action is a step within the activity. An activity diagram is much like a flowchart that shows the flow of control from activity to activity [BRJ99]. Activity diagrams are one of the UML diagrams that are used to model dynamic aspects of systems. Usually they are used to model sequential execution of steps that starts with an initial state and ends with an end state. Activity diagrams can be used to model concurrent execution of steps in a workflow.

An example of an activity diagram is shown in Figure 6-10. An activity is an ongoing nonatomic execution within a state machine [BRJ99]. An activity diagram always starts with a *start state* (or initial state) represented by a filled bullet. Arrows show the transition from one activity to the next one. As an example, in Figure 6-10, when the execution of activity *initialize weather* is terminated, then the execution of the next activity named *initialize soil* takes place.

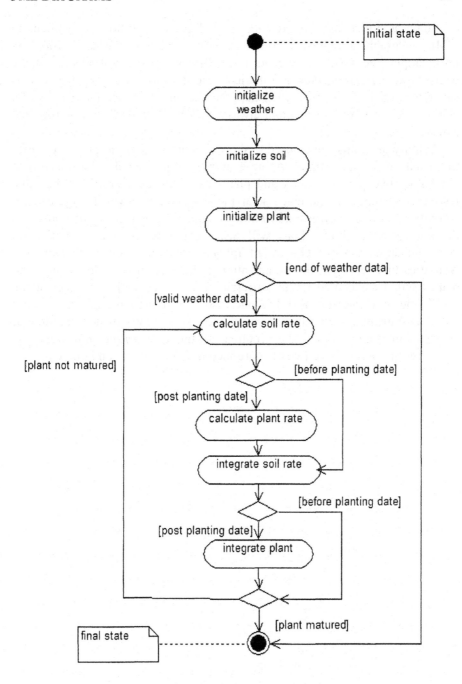

Figure 6-10. Example of an activity diagram.

Some activities will be executed only if some condition is satisfied. In UML, the graphical representation of a condition is referred to as *decision*. As an example, the activity *calculate plant rate* will be executed only if the decision *post planting date* is satisfied. The decision *post planting date* is satisfied if the current date is later than the planting date (i.e., the plant is of a certain age). An activity diagram always ends with an *ending state* (or final state).

Activity diagrams can be used to show concurrent activities, (i.e., activities that occur at the same time). As an example, let us consider the scenario of simulating two plants (plantA and plantB) that are competing for the same resources (water, soil nutrients, solar radiation, etc.) shown in Figure 6-11. Activities *initialize weather* and *initialize soil* are executed sequentially. As the simulation of both plants will occur concurrently, then a horizontal synchronization bar (or concurrent fork) is used to express concurrency. Activities that represent the initialization and the simulation for plantA and plantB will be executed within different flows of control that occur at the same time. A concurrent fork has one incoming transition and two or more outgoing transitions. After the concurrent activities are terminated, the flow of control joins the sequential execution at the *concurrent join point*. A concurrent join may have two or more incoming transitions and one outgoing transition.

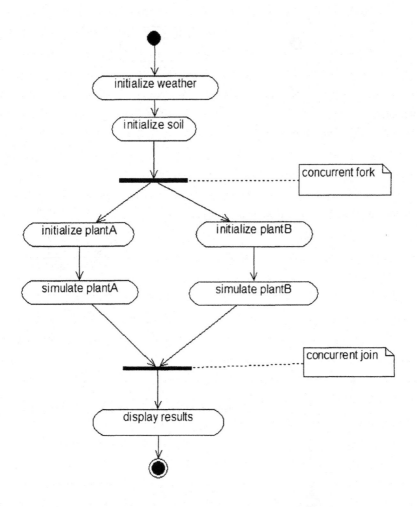

Figure 6-11. Example of concurrent processing of activities.

5. STATECHART DIAGRAMS

Statechart diagrams are one of the five kinds of diagrams UML uses to model dynamic aspects of systems. They are used to model different states of an object during its lifetime; from the time it is created until it is destroyed. A statechart diagram shows the flow of control from one state to another. Figure 6-12 shows an example of a statechart diagram.

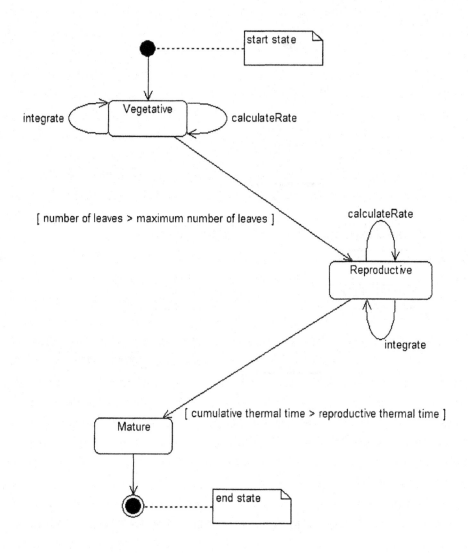

Figure 6-12. States of object plant during the simulation.

A statechart diagram starts with an *initial state* represented by a filled bullet and it ends with an *end state* as shown in Figure 6-12. The statechart diagram represents the different phenological phases (or states) of object *Plant* during the simulation process, described in detail by [PBJ99].

The plant's phenological phase is important as it determines the calculation of plant parameters such as *deltaLeafNumber* calculated by Equation 1.

$$deltaLeafNumber = \begin{cases} temperatureStress \times maxRateOfLeafAppearence \\ \text{when } phenologicalPhase = \text{"vegetative"} \\ \text{and} \\ 0 \\ \text{when } phenologicalPhase = \text{"reproductive"} \end{cases}$$

Equation 1

The diagram shows that at beginning of the simulation (provided that the current simulation date is greater than planting date), the plant is in the phenological phase of vegetation. During that phase, it will receive messages *calculateRate* and *integrate* and will remain in the same phase (or state) as long as the event *number of leaves > maximum number of leaves* does not occur. A state is a condition or situation in the life of an object during which it satisfies some condition, performs some activity, or waits for some event [BRJ99].

When the event *number of leaves > maximum number of leaves* occurs, then object *Plant* will change phenological phase to reproductive. It is important to note that message *integrate* affects parameters such as number of leaves that is used to trigger the event that will send the object plant to the phenological phase of reproductive. An event can trigger a state transition. A transition is a relationship between two states indicating that an object in the first state will perform certain actions and enter the second state when a specified event occurs and specified conditions are satisfied [BRJ99].

During the phenological phase of reproductive, the object plant will receive messages *calculateRate* and *integrate*, and will remain in this phase as long as the event *cumulative thermal time > reproductive thermal time* does not occur. When this event occurs, plant will move to the phase of maturity and this signals the end of the simulation.

Chapter 7

DESIGN PATTERNS

1. A SHORT HISTORY OF DESIGN PATTERNS

Well before software engineers started using patterns, an architect named Christopher Alexander wrote two books that describe the use of patterns in building architecture and urban planning. The first book is titled *A Pattern Language: Towns, Buildings, Construction* [Alex77], published in 1977. The second one is titled *The Timeless Way of Building* [Alex79], published in 1979. These two books not only changed the way structures were built, but they had a significant impact in another not closely related field, the field of software engineering.

According to Alexander, *a pattern describes a problem which occurs over and over again in our environment, and then describes the core of the solution to that problem, in such a way that you can use this solution a million of times over, without ever doing it the same way twice* [Alex77]. Although Alexander refers to buildings and towns, his conclusion can be successfully applied in the process of object-oriented design. Very often programmers have to solve the same problem that occurs in different applications regardless of the problem domain. Saving data in a database, for example, requires the same logic regardless of the amount and the nature of the data used. In the case of object-oriented databases, in order to read an object from the database, the following operations need to be accomplished:

A database session needs to be created in order to access any data in the database.

- Within the session, a transaction should be opened

- Within the transaction, object is read from the database
- If the transaction fails, a rollback to initial values occurs
- If the transaction succeeds, a commit occurs
- Transaction is closed
- Session is closed

These operations are repeated again and again every time an object is read or stored in the database. Therefore, they can be considered as a pattern that can be used by any programmer that needs to communicate with a database. The above list of operations is tested and proved to be correct. A novice programmer does not have to reinvent everything by himself; he can apply the *read pattern* and obtain the right results.

The first work on design patterns was undertaken by Cunnigham and Beck [CK87]. They presented five patterns for user interface design. In mid 90s, a group of four software engineers [GHJ95] wrote the book titled *Design Patterns* that had a significant impact in the way software design was carried out. The book presents well-thought solutions for a large class of problems. The same way an architect uses prefabricated blocks for building complex constructions, a programmer will use patterns to develop complex software. The concept of design patterns allows novice programmers to use elegant solutions provided by experts. Using patterns makes the process of designing complex systems easier.

Design patterns are divided in three categories: Creational, structural and behavioral patterns [GHJ95]. Some other authors such as Grand [Gra98] have created an additional group referred to as fundamental patterns where they include patterns that are used by other patterns. This book follows Grand's classification and starts the presentation of design patterns with fundamental patterns.

Creational patterns deal with the process of creating objects. They describe optimal ways of creating new objects. Structural patterns describe how to compose classes or objects. Behavioral patterns describe how to distribute behavior among classes and how classes interact with each other.

2.   FUNDAMENTAL DESIGN PATTERNS

2.1   The delegation pattern

The purpose of the delegation pattern is to extend and reuse the functionality of a class by writing an additional class with added functionality that uses instances of the first class to provide the original behavior [Gra98].

Often, the reuse of the behavior of a class is realized through the mechanism of inheritance; a subclass inherits from its superclass data and behavior. Inheritance allows classes to be defined based on existing ones. When a new class of objects is defined, only the properties that will differ from the properties of the existing class need to be defined. Other properties defined in the existing class will be included in the new class definition. Therefore, inheritance is considered a mechanism for incremental programming. Code can be reused simply by inheriting it. Inheritance is a static relationship. When a class subclasses another one, their relationship is static and does not change over time.

Inheritance should be used only in the cases when the created subclass is a kind of the superclass, meaning that the subclass and the superclass are conceptually the same. The subclass should not radically alter the behavior of the superclass.

In cases when the existing behavior of a class needs to be extended and the result class is not conceptually similar to the superclass, inheritance should not be used. Another form of reuse, referred to as delegation, is the appropriate way to extend the behavior of a class as shown in Figure 7-1.

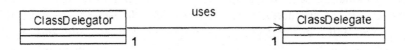

Figure 7-1. Example of delegation.

Figure 7-1 shows that class *ClassDelegator* is associated with class *ClassDelegate* through a unidirectional association. Therefore, *ClassDelegator* has access to data and behavior of *ClassDelegate*; an attribute of type *ClassDelegate* is defined in the *ClassDelegator* that points to *ClassDelegate*. Thus, the behavior of *ClassDelegator* can be extended by using the behavior of *ClassDelegate*.

Let us consider a concrete example of using the pattern of delegation. Figure 7-2 shows the relationships between classes in simple *Plant* component.

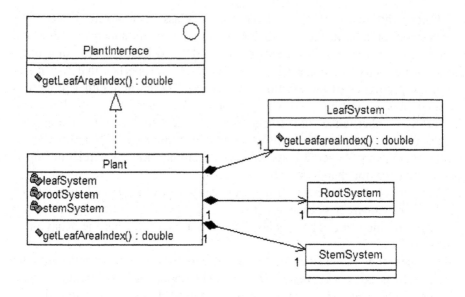

Figure 7-2. Example of delegation.

As explained in section **Components,** the functionalities provided by the component *Plant* are defined by interface *PlantInterface*. To make things simple, we will consider that *PlantInterface* defines only one method, *getLeafAreaIndex()*. As shown in Figure 7-2, *Plant* component is composed of a few classes, such as *LeafSystem, RootSystem,* and *StemSystem*. These classes are provided with data and behavior to play the role of leaf, stem, and root systems of the plant. Although these classes have a well-defined role, none of them communicates directly with other classes or components of the system. In the case that some other class/component of the system would require to use behavior defined in class *LeafSystem*, the communication will occur through class *Plant*. *Plant* has access to data and behavior of classes *LeafSystem, RootSystem,* and *StemSystem*, through attributes *leafSystem, rootSystem,* and *stemSystem* that hold references to objects of the corresponding classes.

When another class or component needs to use the value of leaf area index parameter that is stored in an object of class *LeafSystem*, it needs to send the message *getLeafAreaIndex()* to an object of type *Plant*. This object knows how to respond to this message as it implements *PlantInterface*. Or the calculations for the leaf area index parameter are defined in *LeafSystem*, not in *Plant*. *Plant* will delegate the call to *LeafSystem* using the attribute *leafSystem*. After the calculations are terminated, *Plant* will return the results to the requestor object.

Figure 7-3 shows the implementation in Java of the delegation pattern. Lines 3, 4, and 5 define attributes of *Plant* that point to objects of type *LeafSystem, RootSystem* and *StemSystem*. Lines 7 through 9 define method *getLeafAreaIndex()*. Class *Plant* provides an implementation for this method as it implements *PlantInterface*. As shown in line 8, *Plant* delegates the method call to *LeafSystem,* meaning that method *getLeafAreaIndex()* defined in *Plant* will return the result of the execution of the same method defined in class *LeafSystem*.

```
1   public class Plant implements PlantInterface {
2
3       LeafSystem leafSystem = new LeafSystem();
4       RootSystem rootSystem = new RootSystem();
5       StemSystem stemSystem = new StemSystem();
6
7       public double getLeafAreaIndex() {
8           return leafSystem.getLeafAreaIndex();
9       }
10  }
```

Figure 7-3. Implementation in Java of the delegation pattern.

Using the same principle of delegation, class *Plant* will be able to respond to messages that are defined in classes *RootSystem* and *StemSystem*. Class *Plant* is the main distributing hub for component *Plant*, as it controls access to other classes. If *RootSystem* and *StemSystem* classes had a direct link with classes outside *Plant* component, that would have compromised the principle of encapsulation that needs to be observed during the component design process.

3.   CREATIONAL PATTERNS

3.1   The factory method pattern

The object-oriented approach is about creating objects with a specific behavior to interact with other objects. It is a good modeling practice to make the process of creating objects localized, so if changes have to be made to the way objects are created, these changes will occur in only one place in the code. The factory method pattern is about organizing the object creation process in such a way that new type of objects can be added to the system without reconsidering what is already in place. This is reached by forcing the process of object creation to occur through a common factory, rather then allowing it to be dispersed all over the application. If a new type of object needs to be added to the system, then appropriate changes need to be made to the factory that creates objects.

Let us consider a crop simulation scenario in which many instances of plants will be created such as maize, rice, and wheat. All of these types of plants have some attributes in common such as variety, planting date, etc. The common data and behavior can be defined in an abstract class referred to as *Plant* and subclasses of *Plant* will define additional data and behavior to describe a particular type of plant. The crop simulation system should be able to simulate many types of crops; therefore, objects of different plants will be created. As mentioned before, one solution can be that the process of object creation can be dispersed throughout the code and objects are created when and where they are needed. This solution has a big disadvantage. When changes are needed to be made to the process of object creation, we will need to find all the places where objects are created and do the necessary changes.

The factory method pattern provides a better way of creating objects and works as shown in Figure 7-4.

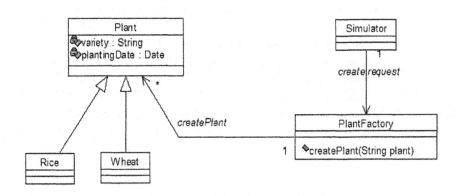

Figure 7-4. The factory method pattern.

As shown in this figure, the *Simulator* sends a *create* request to *PlantFactory* class to create an object of a particular type of plant. *PlantFactory* is provided with a method referred to as *createPlant(String plant)* used to create the required object. This method has a parameter of type String named *plant* that indicates the type of the object to be created. Thus, if an object of type *Maize* needs to be created, then the *Simulator* sends the message *createPlant (maize)* to *PlantFactory*. The following code shows the implementation in Java of the class *PlantFactory*. The kinds of objects that need to be created need to be known in advance and the corresponding code is included in the factory. Figure 7-5 shows that the following factory creates objects of type *Maize, Wheat,* and *Rice.* If new kinds of objects need to be created, then the factory needs to be updated but the update will occur in one and well-defined place.

```
1      package FactoryMethodPattern;
2
3        public class PlantFactory  {
4          public PlantFactory(){}
5
6          public Plant createPlant (String plantName)  {
7
8            Plant newPlant = null;
9            if (plantName.equals("maize"))   newPlant = new Maize();
10           else
11             if (plantName.equals("wheat")) newPlant=new Wheat();
12             else
13               if (plantName.equals("rice")) newPlant = new Rice();
14
15           return newPlant;
16         }
17     }
```

Figure 7-5. Implementation of class PlantFactory in Java.

As shown in Figure 7-5, line 1 shows the package (subdirectory) where the class *PlantFactory* is stored. Line 3 shows the class definition. Line 4 shows the default constructor of class *PlantFactory*. Lines 6 through 16 show that the definition of the method *createPlant*(String *plantName*) considers three cases of creation of new objects: Maize, rice and wheat. The newly created object depends on the value of the parameter *plantName*. Line 15 returns the newly created object. If additional types of plants need to be created, then there is only one place to make the corresponding changes; the method *createPlant* of class *PlantFactory*.

3.2   The abstract factory pattern

The intent of this pattern is to provide an interface for creating families of related or dependent objects without specifying their concrete classes [GHJ95].

To better understand how this pattern works, let us consider the following example. Suppose that the heart of a crop simulation system is an object provided with supervising behavior that controls the objects that need to be created during the simulation process. We will refer to this object as Simulator. Suppose that any plant, be it perennial or annual, is represented by its *LeafSystem* and *RootSystem*; a plant is conceived as a composition of these composing parts. It will be highly desirable to develop a simulation system that is built in a very generic way and capable to simulate different types of plants: Perennials or annuals. Such a system will be independent of the specific plant that is simulating. Therefore, *Simulator* will be asked to create instances (and their composing parts) of perennial or annual plants. How can we develop a generic simulation system that handles any type of plants? The abstract factory pattern can be used to solve this type of issues. The class diagram for the abstract factory pattern is shown in Figure 7-6.

AbstractPlantFactory is an abstract class that provides the most common behavior needed for simulating perennial and annual plants. The behavior needed to simulate a specific plant will be provided by a specific "factory," that is, subclass of *AbstractPlantFactory*. *PerennialFactory* is used to create composing parts of a perennial plant and *AnnualFactory* is used to create the composing parts of an annual plant. *AbstractPlantFactory* is provided with a method referred to as *getFactory* that will deliver the required factory. Once Simulator knows the kind of plant it needs to simulate, it will use the *getFactory* method to obtain the right factory.

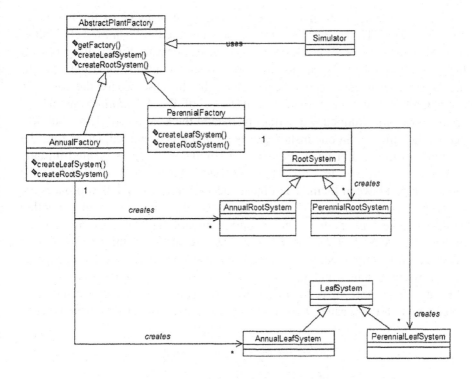

Figure 7-6. Class diagram of the abstract factory pattern.

Note that both factories are provided with behavior capable of creating instances of *LeafSystem* and *RootSystem* for both types of plants. The use of the abstract factory pattern makes it easy to add new types of factories for creating new types of objects by only adding a few lines of code in a well-defined place. It is important to note that the user of the abstract factory, *Simulator* in this case, it completely independent of the factory system that creates objects.

3.3   The singleton pattern

The purpose of the singleton pattern is to ensure that only one instance of a specific class can be created [GHJ95]. In the case where an instance is created from a class defined using the singleton pattern, all other objects of the system that need to dialog with it will refer to the same instance.

Often there are cases that only one instance of a class need to be created. This is common in cases when a class needs to control and coordinate the behavior of other classes. As an example, when designing a crop simulation model, it is important to make sure that only one instance of the *Simulator* class is created. This is because of the particular role the *Simulator* class plays in the simulation process. The *Simulator* is responsible for creating objects that are involved in the simulation process and controls the time and the messages that objects send to each other. In the case where the user can accidentally create more than one instance of the *Simulator*, unexpected results could be obtained. To better understand the kind of problems that may occur when more than one instance of the *Simulator* is created, let us consider a simple example. Let us suppose that a simulator object is created initially and this object creates all other needed objects and establishes the relationships between them. The objects needed in the simulation process are of type *Soil*, *Plant*, and *Weather*. Suppose that at some point, another simulator object is created. The second simulator will have to create other objects of type *Soil*, *Plant* and *Weather*. As a result of having several objects of type *Soil*, it is not clear which object *Soil* is used at a time. *Soil* objects may be in different states and therefore, holding different values for the same attribute. Therefore, it is important to make sure that only one instance of object simulator is created. Figure 7-7 shows an example of singleton pattern.

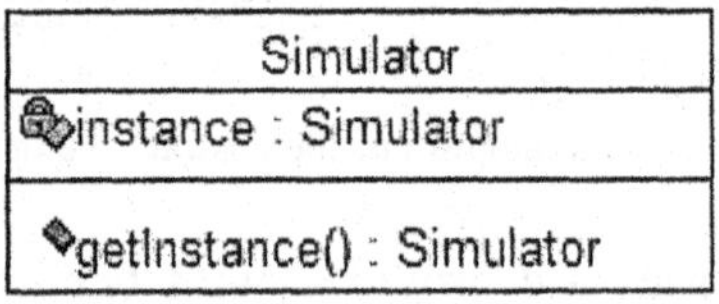

Figure 7-7. Example of implementation of the singleton pattern.

The key to designing a singleton is to not allow users to create more than one instance of the class. Therefore, the attribute *instance* that references the unique instance of the class *Simulator* is defined as private, meaning that this attribute cannot be accessed from outside the class *Simulator*. A method, referred to as *getInstance()*, returns the value of the unique instance. Figure 7-8 shows the Java implementation of the singleton pattern.

```
1    package SingletonPattern;
2
3    final class Simulator {
4        private static Simulator instance = null;
```

Figure 7-8. Java implementation of the singleton pattern (Part 1 of 2).

```
5
6              private Simulator(){};
7              public static Simulator getInstance()  {
8                  if (instance = null) return  new Simulator();
9                  else return instance;
10           }
11   }
```

Figure 7-8. Java implementation of the singleton pattern (Part 2 of 2).

Line 1 shows the subdirectory where the class is stored. Line 3 defines class *Simulator* as final. By making the class *Simulator* final, it prevents cloneability to be added to this class by the mechanism of inheritance. The Java environment allows object creation through cloning. As *Simulator* is a subclass of the superclass *Object*, the *clone* method defined in *Object* normally is inherited and becomes part of the behavior of *Simulator*. Or making class *Simulator* final prohibits the inheritance mechanism from occurring. Thus, objects of class *Simulator* cannot be cloned. This is another step to make sure that class *Simulator* will not be able to create more than one instance of it. Line 4 assigns attribute *instance* to null. Line 6 defines a private constructor to prevent the compiler from inserting a default constructor. Lines 7 through 10 define the method *getInstance()* that delivers the only instance of the class Simulator. In the case where no instance of Simulator is yet created, then line 8 will create it. In the case where the unique instance is already created, line 9 simply returns this instance. This way of creating an instance of an object only when it is needed is referred to as **lazy initialization**.

4.   STRUCTURAL PATTERNS

4.1   The adaptor pattern

The purpose of the adaptor pattern is to convert the interface of a class into another interface clients expect [GHJ95]. The adaptor makes it possible for two classes to work together when they cannot communicate with each other because they implement different interfaces.

To better understand the need for the adaptor pattern, let us take a closer look at two different water-balance models described by [PSh04]. The models are Ritchie's water-balance model [Rit98] and the ISM (Irrigation Scheduling Model) developed by [GSR00].

Ritchie's model uses the United States Soil Conservation Service (SCS) method to determine runoff and in turn, calculate the amount of water that enters the soil surface. This method in Ritchie's model is referred to as *calculateInfiltration*. The ISM model describes the amount of water entering the soil as effective rainfall and uses the SCS method to determine this amount. In the ISM model, the same method is referred to as *calculateEffectiveRainfallSCS*. Both models refer to the same process using different names.

Let us suppose that each of these models is developed as a component that can be plugged into some decision support system used for crop yield simulation. Ritchie's component will provide the amount of water that enters the soil surface by responding to the message *calculateInfiltration*, whereas the ISM component will provide the same result by responding to the message *calculateEffectiveRainfallSCS*. Let us suppose that initially the decision support system uses Ritchie's component. The object that needs to know the amount of water entering the soil sends the appropriate message to Ritchie's component. In the case that there is a need to replace Ritchie's component with the ISM component, there is a problem as Ritchie's and ISM components implement different interfaces. An object that can communicate with Ritchie's component by sending the message *calculateInfiltration* cannot communicate with the ISM component as the latter does not recognize this message. The sender of the message *calculateInfiltration* ignores the fact that the ISM component does not understand this message and that the understood message is *calculateEffectiveRainfallSCS*. Although the developers have used good design techniques to develop these two models as components, the reusability of components is impacted by the different naming conventions used by different authors.

This problem can be solved using the adaptor pattern as shown in Figure 7-9.

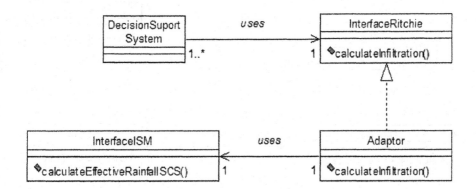

Figure 7-9. Class diagram for the adaptor pattern.

The decision support system uses Ritchie's model. This is shown in Figure 7-9 by the association *uses* between the *DecisionSupportSystem* class and the *InterfaceRitchie* interface. *InterfaceRitchie* defines a method referred to as *calculateInfiltration*. The *Adaptor* class implements *InterfaceRitchie*; therefore, it should provide an implementation for the method *calculateInfiltration*. Class *Adaptor* has an association with *InterfaceISM*, which allows *Adaptor* to access data and behavior from *InterfaceISM*. The body of the method *calculateInfiltration* defined in the *Adaptor* class does not do calculations, but it simply delegates execution to the method *calculateEffectiveRainfallSCS* defined in *InterfaceISM*.

As the *DecisionSupportSystem* class knows how to call the method *calculateInfiltration* from *InterfaceRitchie*, it therefore knows how to call the same method from class *Adaptor* because *Adaptor* implements *InterfaceRitchie*. Therefore, the *DecisionSupportSystem* class can call a method defined in *InterfaceISM* even when it cannot communicate directly with this interface. The *Adaptor* class makes possible the communication between two classes when there is no association between them, as shown in Figure 7-10.

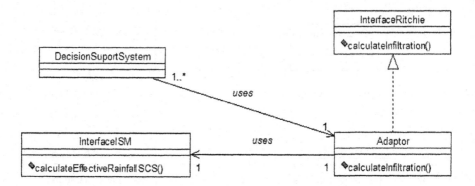

Figure 7-10. An adaptor allows communication between classes that do not implement the same interface.

Figures 7-11, 7-12, and 7-13 are examples of a simple implementation in Java of classes that are involved in the adaptor pattern presented in Figure 7-9. Figure 7-11 shows a simplified definition of *InterfaceISM* that defines the public interface for all the classes implementing this interface. This interface defines a method referred to as *calculateEffectiveRainfallSCS* that will calculate the amount of water that enters the soil surface. Interfaces only define the name of the methods and their signature; they do not provide the logic of the implementation.

```
1    package AdaptorPattern;
2      public interface InterfaceISM {
3          public double calculateEffectiveRainfallSCS();
4      }
```

Figure 7-11. Definition of InterfaceISM.

Figure 7-12 shows a simplified definition of the interface *InterfaceRitchie* that defines the public interface of all the classes implementing this interface. This interface also, defines a method referred to as *calculateInfiltration* that calculates the amount of water entering the soil surface according to Ritchie's model. Figure 7-13 shows the definition of class *Adaptor*. Line 3 shows that class *Adaptor* implements *InterfaceRitchie*; therefore, *Adaptor* should provide an implementation of the method *calculateInfiltration* defined by the interface. Line 5 shows that *Adaptor* has an attribute of type *InterfaceISM* that allows class *Adaptor* to access data and behavior from any class implementing *InterfaceISM*. Lines 7 to 10 show the definition of method *calculateInfiltration*. The result of this method is of type double. Line 9 shows that the execution is delegated to object *ism* that points to an object of type *InterfaceISM*. The result of the method *calculateInfiltration* is obtained by executing the method *calculateEffectiveRainfallSCS* defined in *InterfaceISM*. In this way, object *DecisionSupportSystem* can send the message *calculateInfiltration* to object *InterfaceISM* without knowing about the latter.

```
1   package AdaptorPattern;
2     public interface InterfaceRitchie  {
3       public double calculateInfiltration();
4   }
```

Figure 7-12. Definition of InterfaceRitchie.

```
1   package AdaptorPattern;
2     public class Adaptor implements InterfaceRitchie  {
3       private InterfaceISM ism;
4       public double calculateInfiltration() {
5         return ism.calculateEffectiveRainfallSCS();
6       }
7   }
```

Figure 7-13. Definition of class Adaptor.

4.2   The proxy pattern

The intent for this pattern is to add a level of indirection with a surrogate object that provides the same services as the real object. The surrogate object is responsible for controlling access to the real object [Lar02].

This pattern is almost never used by itself; usually it is used by other patterns. The proxy pattern plays an important role in middleware software such as Java's RMI and CORBA. These technologies are presented later in this book.

A proxy object is designed to receive calls for another real object, the object that provides the required service. Therefore, the proxy and the real object should provide the same services. The proxy object does not provide any implementation for the behavior defined in the real object; it only delegates the call to this one. Thus, the proxy object makes the location of the real object irrelevant to its clients. A proxy object is located near the client that needs the services. Clients that use the behavior of the real object are not aware of the location of this object; the proxy makes it look as clients are communicating with the real object. Figure 7-14 shows class diagrams for the proxy object.

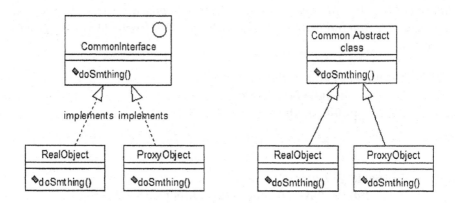

Figure 7-14. Class diagram for the proxy object.

As shown in this figure, the real object and the proxy object may implement the same interface or they may have a common superclass. In the case they implement the same interface, the real object and the proxy object provide a polymorphic implementation of the behavior defined in the interface. While the real object provides an implementation for the behavior defined in the interface, the proxy's behavior is to delegate the call to the real object.

How do the proxy and the real object communicate with each other? Figure 7-15 shows the communication between the proxy and the real object.

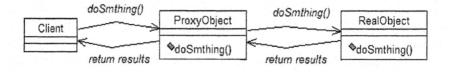

Figure 7-15. Communication between the proxy and the real object.

A client that needs to use the services provided by the real object sends a message to the proxy. As the proxy object implements the same interface as the real object, it recognizes the method call and it delegates it to the real object. The real object executes the call and returns the results to the proxy object that communicates them to the client. The client is unaware of the fact that the service requested is provided through the proxy. More details about the implementation of the proxy object will be provided later in this book in Chapter 11.

4.3   The iterator pattern

The purpose of the iterator pattern is to separate the logic of an algorithm manipulating the data from the particular structure of the container containing the data [GHJ95].

As an example, let us consider the problem of obtaining weather data for a simulation model. As previously mentioned, different authors have solved this problem using different mechanisms; some authors read the weather data from a text file saved locally in the system [HWH01]. Others have developed complex systems to obtain weather data from networks of real-time weather stations [LKN02].

Algorithms used in both cases have things in common and things that are different. The things in common are that in both cases an iteration is used to sequentially analyze each of the daily (or other time unit used) data. Things that are different are the particular data structures or data containers used in each of the cases. In the case that data are saved in a text file, the data container is a file containing lines of data, each line containing weather data for a day or other time unit used in the simulation. At each step of the iteration, a line containing the daily data will be read and the corresponding values will be assigned to variables designed to hold them. In the case that weather data are obtained from a weather station, the data container can be a table that is returned from the execution of an SQL statement. Each row of the table represents daily data. Looping through the container, an object holding the daily weather data will be returned and each of the daily weather parameters can be obtained by sending the appropriate message to this object. As an example, to get the rainfall for the day, the message *getRainfall* should be sent to the object holding the daily data.

It is desirable to design the *Weather* object in a general way that multiple sources could be used and independently of the particular container used to hold the data. This problem can be solved using the *Iterator* pattern. Figure 7-16 shows a class diagram for classes involved in the *Iterator* pattern.

The class diagram shown in this figure is taken from a crop simulation scenario. Object *Simulator* needs daily weather data for the simulation process. The *Simulator* has access to *WeatherInterface* that defines the operations needed to obtain the weather data. *WeatherInterface* implements interface *Iterator* provided by the Java programming environment that makes available the logic for iterating over a data container. Therefore,

WeatherInterface defines iterator behavior as well. Class *Weather* implements the *WeatherInterface*; therefore, it will provide the behavior necessary for obtaining the data from a particular data container. Class *Weather* has access to *WeatherDataContainer* from where it will extract the data. *WeatherDataContainer* is a collection of *DailyWeatherData* that is an object with attributes such as rainfall, minTemperature, maxTemperature, solarRadiation, etc. The *Simulator* has access to an object of type *DailyWeatherData* and can obtain the values of rainfall, temperature and solar radiation by sending to this object messages such as *getRainFall*, *getSolarRadiation*, etc.

It is important to note that the *Simulator* that requests the data is completely independent of the class *Weather* that provides them. As *Simulator* has access to the interface *WeatherInterface*, not to the concrete class *Weather*, it can use any *Weather* object that provides a polymorphic implementation of this interface. Hence, this architecture has two advantages: First decouples *Simulator* from the concrete class *Weather* and second, it makes the algorithm that uses the weather data independent from the particular data container. Therefore, different sources of weather data can be used in the simulation process.

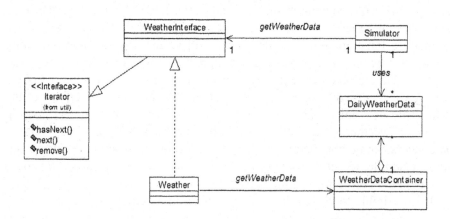

Figure 7-16. Classes involved in the iterator pattern.

Figure 7-17 shows another way of imposing class *Weather* to provide *Iterator* type of behavior. Class *Weather* implements two interfaces and will provide the behavior defined in both interfaces.

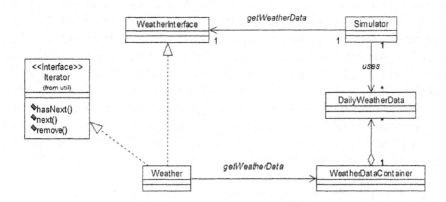

Figure 7-17. Another class diagram for Iterator.

A detailed implementation of this pattern in Java is provided in section **Implementation of the Kraalingen model in Java** of Chapter 8.

4.4   The façade pattern

The purpose of the façade pattern is to provide an interface to a set of interfaces in a subsystem [GHJ95]. Using this pattern makes the objects included in the subsystem easier to use.

To explain the context in which the façade pattern can be used, let us refer again to an application that is based in a crop simulation model. The application will have a graphical user interface (GUI) that will allow users to enter the required initial data. The heart of the application is the crop simulation model that will run the simulation using the entries provided by the user. Most of the crop simulation models use a core of objects or components such as *Weather, Soil, Plant,* and *SoilPlantAtmosphere.* The simulation results may be used for purposes such as to display different graphics or maps or to create reports or other operations that are needed to satisfy user's requirements. The communication between the user interface and the crop simulation model is bidirectional; users will enter initial data and the simulation model will return the results of the simulation to the user's interface.

When environmental-based applications are developed, the most common approach is to strongly link the code needed for the user interface with the code representing the environmental model. This approach may allow for fast software development, as any object can be accessed by any other object anywhere, but it is not an optimal one, as it creates long-term problems related to code maintenance and reuse. Although the user interface and the environmental model exchange data between them and are part of the same application, they represent two separate parts of the system and therefore should be designed to function independently. Having strongly coupled the user interface with objects/components such as *Soil*, *Weather*, and *Plant* makes the system hard to maintain and difficult to reuse. In such a highly coupled system, no part of it can be reused or easily modified.

This kind of problem can be solved using the façade pattern. The essence of this pattern is to provide a unique point of communication between an object and the surrounding environment. Figure 7-18 shows the class diagram for the façade pattern in general.

As shown in this figure, communication between client objects and a group of other objects is filtered by the *FacadeClass*. Therefore, *FacadeClass* should be provided with the right behavior in order to represent the rest of the objects in communication with the surrounding environment.

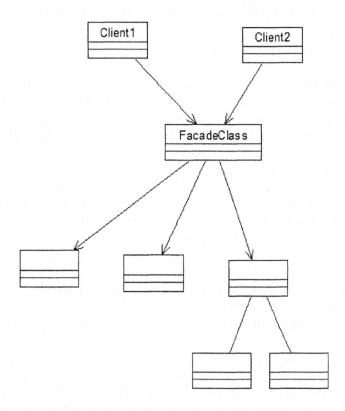

Figure 7-18. Class diagram for the facade pattern.

Figure 7-19 shows the class diagram for the crop simulation example. It is important to note that the associations between *Simulator* and the GUI classes are bidirectional. *Simulator* controls all the communication between the GUI classes and crop simulation objects.

Although the *FacadeClass* controls the communication between clients and a group of classes implementing some abstraction, it is not necessary that the *FacadeClass* be a rigorous barrier between them. In some cases it is advised to let clients have access to particular objects behind the façade. In these cases, the *FacadeClass* should be provided with the behavior that makes its objects accessible to clients. Figure 7-19 shows that *Simulator* is provided with operations that can make available to GUI objects any of the *Plant, Weather, Soil,* or *PlantSoilAtmosphere* objects.

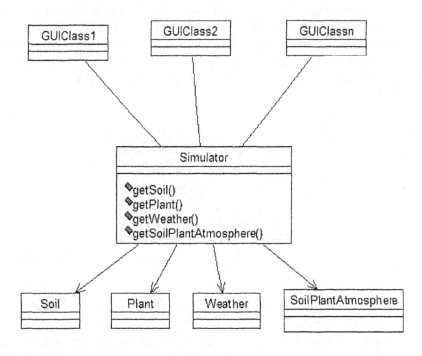

Figure 7-19. Communication with GUI classes goes through the Simulator.

5.   BEHAVIORAL PATTERNS

5.1   The state pattern

The purpose of the state pattern is to allow an object to alter its behavior when its internal state changes [GHJ95]. Using this pattern makes it easier to model some complicated scenarios often present in biological simulation models.

Let us consider the example of the lifecycle of an insect. At the beginning, an insect exists in the form of eggs. Later on, as the incubation period ends, eggs are transformed into larvae and finally a larva is transformed into an adult insect. During its different development stages, an insect has different characteristics or properties. An egg has a different behavior from a larva and a larva behaves differently from an adult insect. How can an entity be modeled as an object when it has different characteristics and behavior during different phases of its life? The state pattern can be used to solve this type of problems. Figure 7-20 presents a class diagram for the state pattern.

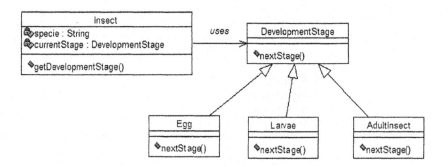

Figure 7-20. The UML diagram for the state pattern.

According to Figure 7-20, class *Insect* has an association with another class named *DevelopmentStage*. Class *DevelopmentStage* is subclassed by three other classes *Egg*, *Larvae*, and *AdultInsect*, representing the three different development stages of an insect. The role of class *DevelopmentStage* is to define an interface common to all subclasses and can be modeled as an abstract class. Class *Insect* defines the common characteristics of the insect regardless of its current development stage such as name of the specie, etc. Because of the relationship with *DevelopmentStage*, class *Insect* has an attribute of type *DevelopmentStage* referred to as *currentStage* that allows access to data and behavior of each of the subclasses. Therefore, class *Insect* can delegate all received messages to the particular subclass that defines the current development stage. It is important to note that attribute *currentStage* can hold only one value at a time, representing the fact that an insect can be at one development stage at the time. If we need to know the current development stage of the insect, then the message *getDevelopmentStage* should be sent to object *Insect* and the result is the name of the class referenced by attribute *currentStage*. Each of the subclasses will define data and behavior necessary to describe the particular development stage of the insect.

Class *DevelopmentStage* has a method referred to as *nextStage*. The purpose of this method is to determine the insect's next stage. As class *DevelopmentStage* is an abstract class, the method *nextStage* will have no implementation. This method is inherited by all the subclasses (*Egg, Larvae,* and *AdultInsect*) and each of the subclasses will provide a specific implementation of this method. As an example, the method *nextStage* applied to an object of class *Egg* will return an instance of type *Larvae*.

When the message *nextStage* is send to an object, for example, to object *Larvae*, besides delivering an object representing the next development stage, the object should destroy itself. Thus, the creation of an object representing the next development stage is accompanied with the destruction of the previous stage; a new object of type *AdultInsect* is created and the current object of type *Larvae* is destroyed. Every time that a change happens, the name of the newly created object is stored in the attribute *currentStage* of class *Insect*.

5.2   The strategy pattern

The intent for this pattern is to define a family of algorithms, encapsulate each one, and make them interchangeable [GHJ95]. Therefore algorithms can vary independently from the clients that use them.

In order to better understand the context in which the strategy pattern operates, let us consider the example of obtaining weather data for a crop simulation system. We have previously mentioned that these data can be obtained using different sources such as using a text file where the data are saved, reading them from a database system, or using an on-line system of weather stations. A well-thought-out system should provide behavior for using several sources of weather data or, in other words, several strategies should be available to users. In a system developed in a traditional programming language such as FORTRAN, the ability to choose between several options would require the use of complex if-then-else statements.

Furthermore, the use of an if-then-else statement will allow for using only known scenarios. In the case where a new way of obtaining weather data is made available, changes to the code are required. Therefore, traditional programming languages offer rather limited and rigid solutions to this problem.

The object-oriented paradigm solves this problem by offering a flexible and better solution. The behavior for using different sources of weather data will be implemented as different classes; each class should provide weather data from a particular source. Then, the question is, how do we choose the right class between several potential ones? The strategy pattern can be used to solve this type of problems. Figure 7-21 shows classes that are involved in the strategy pattern.

The *WeatherDataManager* class provides the behavior for managing the weather data (i.e., provides the capability of using different sources of weather data.) The *WeatherDataProvider* is an interface that represents the common behavior all classes that provide a particular implementation of this interface should implement. Each class is designed to provide data from a particular source. The *WeatherDataManager* has a unidirectional association with *WeatherDataProvider*. The multiplicity of this association allows one manager to use one or zero weather data provider. Classes *WeatherDataFromFile*, *WeatherDataFromStation* and *WeatherDataFrom Database* provide behavior for extracting data from a particular source of data. These classes implement the same interface, the *WeatherDataProvider* interface; therefore, any one of them can be used to provide the weather data requested by the weather data manager.

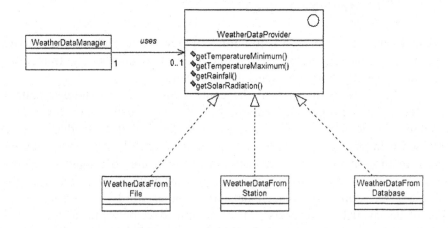

Figure 7-21. Class diagram for the strategy pattern.

Notice that the weather data manager does not have any knowledge of the classes that can provide what the manager wants. The weather data providers are totally independent of the user of the data. This allows for modifying the algorithm that obtains the data without requiring any changes in the user of the data. Furthermore, additional ways of obtaining weather data can be added to the system without forcing the data user to modify its behavior. The user can change the strategy for obtaining weather data without requiring changes to the code. We will see an implementation of this pattern in the second part of the book, in section **Java implementation of the Kraalingen model**.

PART 2: APPLICATIONS

In the first part of the book, we introduced the basic concepts of the object-oriented paradigm and their notations in UML. In the second part of the book, we will see how the knowledge accumulated so far will be used to model a particular problem and develop the corresponding software. We will go through the phases of analysis and design of a simple crop simulation model. The selected model is chosen to be simple on purpose; we would like to avoid getting lost in the details of the crop modeling. Instead, our focus is on the approach used to carry out the analysis and the design using the object-oriented paradigm and construct visual models using UML. The relative simplicity of the selected model does not question the integrity of the used methodology or the nature of the problems encountered and the provided solutions.

Chapter 8 deals with the process of analysis, design, and development of a crop simulation model referred to as the Kraalingen approach [Kra95]. First, a short description of the problem will be provided. Some of the equations used in this model will be presented to demonstrate the links that are needed between model elements such as *Plant*, *Soil*, and *Weather*. Then, the use case shows what the system can offer to users, without showing how these functionalities will be provided. After the use case model is developed, the use case realization is presented for each use case. The use case realization presents several type of diagrams developed to show the dynamic aspects of the system. The diagrams are the sequence and collaboration diagrams, known as interaction diagrams. They help developers to better understand the role and the behavior of each of the potential classes needed to develop the system.

A conceptual model for the Kraalingen approach is presented to show concepts and abstractions from the problem domain and their relationships. The conceptual model shows only one type of class, the classes that represent concepts of the problem. Other classes than the ones that represent concepts are needed; the behavior of these classes is needed to present the graphical user interface (GUI) and the dialog between the user and the system. Finally, the implementation in Java for interfaces and classes used in the system is provided.

Chapter 8

THE KRAALINGEN APPROACH TO CROP SIMULATION

The crop simulation model considered in this study is the one developed by Kraalingen [Kra95]. This approach uses the rate-state concept of simulation modeling [PL82]. Calculations and statements are divided into four categories: Initialization, rate calculations, integration calculations, and the output of results. These calculations are executed sequentially. The simulation starts at the beginning of a time step with a certain value for its state variables; therefore, the initialization step must be performed first. Rate and integration calculations are repeated a certain number of time steps until a termination condition is satisfied. For crop growth, a complete simulation run simulates growth from emergence to harvest. Final calculations and statements are made at the end of a simulation run (e.g., by writing final crop yields to an output file). A detailed description of the model and the FSE (Fortran Simulation Environment) can be found in [BTK00].

Kraalingen has used a modular approach when each module should:

- Read its own parameters;
- Initialize its own variables;
- Accept variables passed to it from other modules and the environment;
- Pass variables that are computed within the module;
- Own its set of state variables;
- Compute rates of change for its state variables;
- Integrate its state variables;
- Write its own variables as output.

In this model, the effect of temperature on daily plant growth is calculated by the equation:

$$PT = 1-0.0025((0.25*Tmin+0.75*Tmax)-26**2 \qquad \textbf{Equation 1}$$

where:

PT = temperature based limiting factor,
$Tmin$ = minimum daily temperature,
$Tmax$ = maximum daily temperature.

The plant cycle is divided in two phases: Vegetative and reproductive. The vegetative phase goes on until the plant reaches a genetically determined maximum leaf number [PBJ99]. In the vegetative phase, the delta leaf area index is calculated by equation:

$$dLAI = SWFAC * PT * PD * EMP1 * dN * (a/(1+a)) \qquad \textbf{Equation 2}$$

where:

$dLAI$ = delta leaf area index,
$SWFAC$ = soil water factor,
PT = temperature-based limiting factor,
PD = plant density,
dN = leaf number increase,
$EMP1$ = empirical coefficient for LAI computation, maximum leaf area expansion per leaf,

and a is calculated by the equation:

$$a = e**(EMP2 * (N - nb)) \qquad\qquad\qquad \textbf{Equation 3}$$

where:

$EMP2$, nb = coefficients in the expolinear equation,
N = plant development stage.

In the vegetative phase, the assimilates are partitioned between canopy and roots whereas in the reproductive phase, all growth occurs in the grain. During the reproductive phase, the difference between daily mean temperature and a base temperature is used to calculate the rate of plant development. Total rate of development towards maturity is accumulated in each step of the simulation [PBJ99].

Our goal is not to make a detailed description of the Kraalingen approach. We are presenting only some of the equations that explain the relationships between simulation elements, plant, soil, and weather. [PBJ99] provide a detailed description of the equations used in this crop simulation model.

1.   SYSTEM REQUIREMENTS

In this section, we will define the requirements of the system. Usually, this part of the project is undertaken in close collaboration with future users of the system. The users should express all their concerns about the future system: The functionalities the system should provide, the way the input data are entered into the system, and the way the final results are presented.

In order to make things simpler, we will consider that the user will need to enter some initial data needed to define the context in which the simulation is running. The initial data are mostly soil and plant data. The initial plant related data are used to populate an instance of class *Plant*; usually the plant initial data are related to the planting date. By entering the planting date as an input parameter, users can study the impact of this parameter on the crop yield. Initial soil-related data are used to populate an instance of class *Soil*; usually the initial soil data are soil depth and wilting point percentage. By providing initial soil data, users can study the impact of these soil parameters on the crop growth. After entering the initial data, the user may start the simulation process.

After performing a simulation, the system will return the results to the user. For reasons of simplicity, we will assume that the results of the simulation can be displayed in the same window as the input data. Therefore, the user will have to use only one window for entering initial data and for displaying the results.

2.   THE USE CASE MODEL

As mentioned in Chapter 5, where we talked about use cases, the use case model represents what the system can do for the users, without explaining how the system will do it. The users of our future system require that the system provide capabilities to enter initial data and perform a simulation. As a first approach we will consider as the use case model the one presented by Figure 8-1. As shown in this figure, users can use the system to enter initial data and to start a simulation.

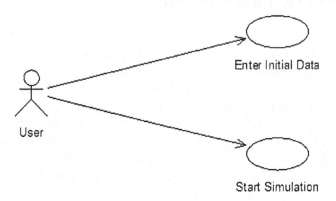

Figure 8-1. The use case model for the Kraalingen approach.

Let us take a closer look at the use case model in Figure 8.1. The use case *Enter Initial Data* represents the set of operations that the user should perform to create initial conditions for the simulation process. The process of entering the initial data is not an independent process that can stand on its own; it is closely related to the process of starting a simulation. The user will not obtain interesting results by performing a simulation with default values, without entering the required initial data. Presenting the set of operations needed to enter the initial data as a separate use case does not match well with the definition of the use case. The use case definition states that the set of operations represented by the use case should have a well-defined purpose and a useful result. Therefore, as the process of entering initial data is closely related to the process of performing a simulation. We will present both activities as one unique use case, as shown in Figure 8-2.

Figure 8-2. Both activities as one unique use case.

2.1   The use case description

After presenting the use case model, it is important to provide a description of what the use case is supposed to do. A use case description helps people involved in the process of the software development to understand the functionalities encapsulated in the use case and to facilitate the discussion about the validity of the use case.

Let us consider the *Start Simulation* use case and provide a potential description for it. One possible brief description of this use case can be the following:

This use case describes how a user can perform a crop simulation process. The data entered by the user define the initial conditions for plant and soil. After the calculations are terminated, the results are displayed in the window.

The brief description is important and if the use case is simple enough, all it can be provided with is the brief description. When the use case is complex, additional information is required. A more detailed description of the use case can be given in the form of an outline. The outline shows the simple steps of the use case using short sentences and presented in a timely manner. The focus at this point is on the clarification of the basic flow of the events. The basic flow (or the main flow) does represent a description of the normal and expected path of the execution of the use case. Some authors refer to the basic flow of events as the happy scenario, where everything goes well and nothing goes wrong.

Later, the focus will shift into presenting the most significant alternatives and exceptions, thus shedding more light on the complexity of the use case. An alternative flow of events represents a possible execution route from the starting point to the end of the use case that is different from the basic flow. An alternative flow represents one of the scenarios where something is not executed as predicted. Thus, most of the alternative flows represent the errors that may occur during the execution of the use case. The basic flow represents the successful route from the beginning to the end of the use case. The alternative flows show all the detours (unsuccessful executions) that may occur during the execution of the use case.

An outline for the use case *Start Simulation* can be the one presented in the next section.

2.2   Basic flow

1. The use case starts when the user clicks on the *Start Simulation* button.
2. Initialize plant, soil, and weather.

3. Loop through the weather data:
   a. Calculate rate for soil and plant.
   b. Integrate soil and plant.
4. End of the weather loop.
5. The use case ends.

2.3   Alternate flow

1. The weather file is not found (in the case that the weather data are provided using a text file stored locally).
2. The communication with the weather station is not possible (in the case that the weather data are provided using an on-line weather station).
   In both cases, the system should stop the execution and display an appropriate error message.

2.4   Preconditions

If we take a closer look at the main flow of events for the use case *Start Simulation*, we realize that the sequence of events may not be meaningful, unless we are at the right starting point of the use case. In the case that the weather data are provided as a text file, the right starting point for the simulation is when the weather file exists and is stored in the right directory. If there is no weather data file, then the simulation cannot be performed. In this case, we can say that the precondition for the use case *Start Simulation* is the following:
A valid weather data file is stored in the right directory of the system.
A precondition is a statement that presents conditions under which the use case can be executed. In the case above, the precondition states that no use case can be executed if there is no weather data file or the file is not placed in the right directory.

2.5   Postconditions

The postcondition is a statement that describes the state of the system when the use case is terminated. The postcondition should be true for all the alternative flows, regardless of which one was executed and it should be false for the basic flow. The idea behind the postcondition is that if anything goes wrong during the execution of the use case, the system should be left in a condition described in the definition of the postcondition. Defining an adequate postcondition is very important as it defines the state the system should be in when the use case terminates.

As an example, in the case when weather data are provided by an on-line weather station one possible postcondition can be defined as follows:

The connection between the user's computer and the server where the weather data are located should be terminated.

The above formulation of the postcondition looks like a very trivial thing, but it is important to know the state of the system when the connection with the weather server fails. Let us suppose that the postcondition for our use case were defined as follows:

The system should continue dialing the weather server until connection is established.

According to this definition of the postcondition, the system will continue dialing the server until the connection is restored. In this case, the system will be busy for as long as the connection is not established. A system designed to persist connecting to the server no matter how long it takes may not be a good system, as the system cannot be used as long as it tries to connect to the remote server.

3. THE USE CASE REALIZATION

In Chapter 5 when we talked about use cases, we pointed out clearly that the use cases only show what the system can provide to its users without explaining how. Therefore, the use case model helps one to understand what the users can ask the system to do, without showing how the system will do it.

Now, it is the time to consider how the system will provide its services to the users. This is achieved by developing for each use case its realization. The realization describes how the behavior of a use case will be provided by collaboration of different elements of the system. The realization of a use case can be presented using UML interaction diagrams or textually using structured English.

It is important to note that the separation of use cases from the use case realization decouples the process of gathering requirements (expressed in a synthetic way in the use case model) from the design of the model (explicitly expressed in the use case realization). This separation allows developers to focus on one well-defined problem at a time and avoid dealing with design issues during the phase of analysis and vice versa. Figure 8-3 shows the UML notation of the use case realization. The dotted eclipse represents the use case realization and the dotted arrow represents the realization association. For each use case of the model, a use case realization should be developed.

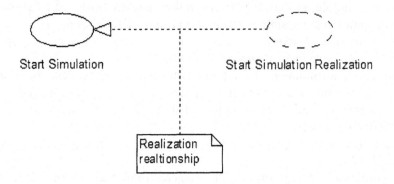

Figure 8-3. The use case realization for Start Simulation use case.

3.1   Sequence diagram for the use case

Figure 8-4 shows the sequence diagram for the *Start Simulation* use case. As shown in this figure, the sequence diagram presents all the elements of the system that participate in the simulation. Messages are numbered to show the order in which they are sent. The process starts with the user sending the message *simulate* to *Simulator*. Messages number 2 through 14 show what the *Simulator* should do in order to fulfill the received request. *Simulator* is responsible for the creation of all needed objects and for sending to each of them the right message at the right time.

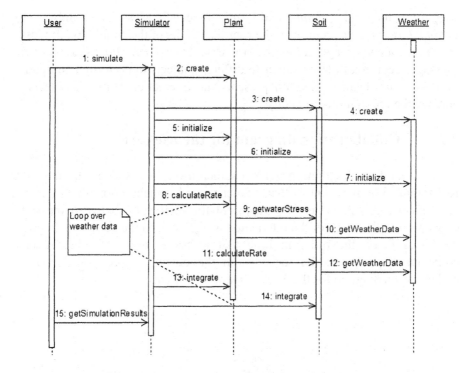

Figure 8-4. Sequence diagram for the Start Simulation use case.

Messages 2, 3, and 4 show that *Simulator* needs to create instances *plant*, *soil*, and *weather* of the corresponding classes *Plant, Soil*, and *Weather*. Messages 5, 6, and 7 initialize each of the instances created in the previous steps. During the initialization process, each of the instances will be populated with initial values, part of which is provided by the user. As an example, the value for the planting date for object *plant* is provided by the user before starting the simulation process. Similarly, initial values for soil depth and wilting point in percent, needed for populating an object *soil*, are provided by the user.

Messages 8 to 14 are part of the iteration over the weather data. These messages will be repeated a certain number of times until the condition plant is mature is satisfied. At the beginning of each iteration, Simulator sends to object plant the message *calculateRate*. In order to calculate the rate object, plant needs soil and weather data. Therefore, object plant sends the message *getWaterStress* to object soil and the message *getWeatherData* to object weather.

The next step in the simulation process is the integration of the values obtained during the rate calculation. Thus, the simulator sends the message *integrate* to both *plant* and *soil* objects. Message number 15, *getSimulationResults*, returns to the user the results of the simulation process.

The sequence diagram for the use case *Start Simulation* shows all the messages sent in a timely manner to different objects to perform a simulation process. This detailed diagram presents the interaction between objects to achieve the required functionality.

3.2   Collaboration diagram for the use case

Collaboration diagrams provide another level of detail of the use case realization. We now know that sequence and collaboration diagrams are semantically the same, as they represent the interaction between the same elements, but the focus of the interaction is different. In the sequence diagram, the focus is on the order in time that messages are sent whereas in a collaboration diagram, the focus is on the object. Figure 8-5 shows the collaboration diagram for the *Start Simulation* use case.

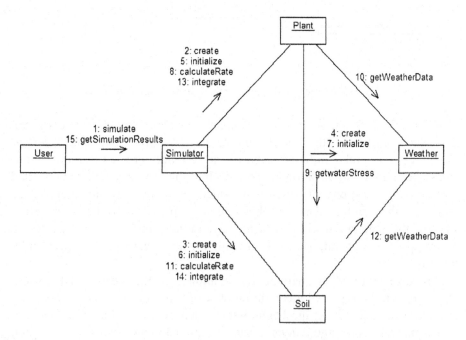

Figure 8-5. Collaboration diagram for the use case Start Simulation.

As shown in Figure 8-5, object *soil* receives messages 3, 6, 11, and 14 from object *simulator*, message 9 from object *plant* and sends message 12 to object *weather*. Therefore, object *soil* should be provided with the appropriate behavior in order to respond to the received messages. Similarly, object *plant* receives messages 2, 5, 8, and 13 from object *simulator* and sends message 9 to object *soil* and message 10 to object *weather*. Therefore, object *plant* should be provided with the appropriate behavior to respond to messages it receives.

As shown in Figure 8-5, a collaboration diagram helps one to understand the kind of behavior objects should provided to be able to successfully dialog with each other to achieve the required functionality. Figure 8-6 shows the behavior defined for class *Soil*, defined by analyzing the collaboration diagram. According to the collaboration diagram, class *Soil* should respond to messages *calculateRate*, *integrate*, *initialize*, and *create*. For the moment, we do not have enough information to define the attributes of the class *Soil*, as we are focused on defining its behavior. Once we know the kind of behavior class *Soil* should provide, then the appropriate attributes will be added to its class definition.

According to the collaboration diagram, *Simulator* receives messages 1 and 15 from the user; therefore, its class definition should include methods named *simulate* and *getSimulationResults*. In the same way, we will define the behavior of class *Plant* as shown in Figure 8-7 and of class *Weather* shown in Figure 8-8.

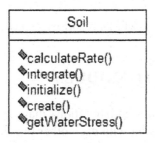

Figure 8-6. The definition of behavior for class Soil.

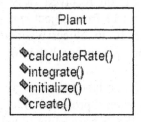

Figure 8-7. The definition of behavior for class Plant.

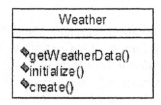

Figure 8-8. The definition of behavior for class Weather.

4.   CONCEPTUAL MODELS

Interaction (sequence and collaboration) diagrams help us to understand how modeling elements dialog with each other to achieve functionality. It is a good modeling practice to start the design process by developing a *conceptual diagram* that represents our knowledge of the application domain expressed through concepts, abstractions, and their relationships. Conceptual diagrams are the result of an activity that is referred to as *conceptual modeling*. Conceptual modeling can be defined as the process of organizing our knowledge of an application domain into hierarchical rankings or orderings of abstractions, in order to obtain a better understanding of the phenomena of concern [Tai96]. Conceptual modeling makes heavy of abstraction and the object-oriented approach, and unlike other programming paradigms, provides direct support for the principle of abstraction. Any entity or concept in a problem domain is conceived as an object provided with a certain state and behavior to play a well-defined role.

Conceptual diagrams have the advantage of presenting the concepts and their relationships in an abstract way, independent of any computing platform or programming language that may be used for their implementation. During this phase, the focus is on depicting the concepts of the system and providing them with the right data and behavior. Experience shows that implementation technologies change constantly. Therefore, it is highly desirable that the model we are about to develop be expressed in an abstract and logical manner resilient to changes.

UML allows for designing a Platform Independent Model (PIM) that presents many advantages. First, PIM allows for representing models using a high level of abstraction. Details of the models can be expressed clearly and precisely in UML as it does not use any particular formalism. UML is semantically very rich, richer than any programming language. The conversion of a UML diagram into code in a particular programming language comes with loss of information. Therefore, the intellectual capital invested in building models will be insulated from changes in the implementation technologies.

After a PIM is developed, then the issue of selecting a particular implementation environment can be addressed. Next, a Platform Specific Model (PSM) will be developed by mapping a PIM to a particular computer platform and a specific programming environment. The transformation of a PIM to a PSM is realized using a mapping process. This two-layer concept, a PIM and a corresponding PSM, separates the scientific model from the implementation technologies. Usually, the science behind the model has a much longer life than the implementation technologies. Changes and evolution of the implementation technologies should not affect the logic of the scientific model. Conceptual diagrams are an important tool for software design. They help to structure the system and a well-structured system is easy to develop, maintain, and reuse. Therefore, it is important to start with a conceptual diagram that presents the core elements and the interactions between them. It is a good modeling practice to name participating elements and their relationships with meaningful names. Meaningful names for concepts and their relationships make the model easier to understand; users can use the conceptual model as a discussion platform where business issues are addressed.

4.1 Conceptual model for the Kraalingen approach

In order to develop a conceptual diagram for the Kraalingen approach, let us take a closer look at the equations of this model. Equations 1, 2, and 3 (see Chapter 8, **The Kraalingen Approach**) represent the relationships between objects *plant*, *soil*, and *weather*. Equation 2 shows that soil water factor data are needed to calculate changes (delta) in the leaf area index of the plant. The same equation shows that the temperature-based limiting factor, calculated by Equation 1, is needed to calculate delta changes in the leaf area index. Equation 2 shows that temperature affects daily plant growth and that the amount of water in soil impacts plant growth as well. Other equations show that plant data are needed for calculations of processes occurring in soil and soil data are needed for calculations occurring in plant. As an example, soil data are needed to calculate daily net photosynthesis processes occurring in the plant. Plant data are needed for calculating potential evapotranspiration, potential soil evaporation, potential plant transpiration, and rate calculation. These are processes that occur in the soil.

Based on the equations expressing relationships between soil, plant, and weather, a diagram for the Kraalingen conceptual model can be presented as shown in Figure 8-9.

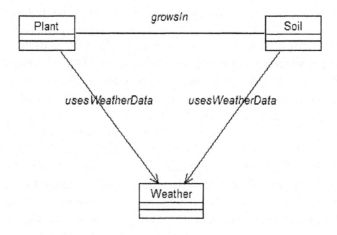

Figure 8-9. Conceptual diagram for the Kraalingen approach.

As shown in Figure 8-9, there is an association referred to as *growsIn* that links *Plant* and *Soil* that reads that plant grows in the soil. This association is bidirectional, meaning that an object of type *Soil* can access data and behavior in an object of type *Plant* and vice versa; an object of type *Plant* can reach data and behavior in an object of type *Soil*.

Weather data are used for calculating soil processes such as potential evapotranspiration, runoff and infiltration. Therefore, the association *usesWeatherData* links *Soil* and *Weather* with navigation direction from *Soil* to *Weather*. Thus, an object of type *Soil* is able to reach data and behavior from an object of type *Weather*. Equation 1 states that weather data are needed to calculate the growth rate reduction factor in plant. Therefore, an association between *Plant* and *Weather* is needed with navigation direction from *Plant* to *Weather*. This association is referred to as *usesWeatherData* and allows an object of type *Plant* to access data and behavior from an object of type *Weather*. The conceptual diagram shows that while objects of type *Plant* and *Soil* should have knowledge of each other, object *Weather* does not have access to any of the objects of type *Plant* and *Soil*. This is because of the particular role the weather data play in the simulation; they are used by other objects to calculate processes occurring in these objects. There are no processes occurring in object *Weather*; therefore, object *Weather* does not need to access data and behavior from objects *Soil* and/or *Plant*.

5.   DISCOVER POTENTIAL CLASSES

The conceptual model shows the interaction between classes that represent concepts from the problem domain; *Weather*, *Plant*, and *Soil* are classes that represent concepts in a crop simulation domain. Although discovering all needed concepts is not an easy task, this is only part of what needs to be achieved. The conceptual model does not show any aspect of the user's interaction with the system. This behavior, how users will interact with the system, will be provided by some other classes that do not represent any of the concepts of the problem domain and therefore, they are not part of the conceptual model.

The conceptual model shows that class *Plant* needs to access data and behavior from classes *Weather* and *Soil*, to calculate processes that occur in class *Plant*. Associations between classes are "communication channels," which allow objects created from these classes, to send/receive messages to one another. Although all the required structures are in place to make the dialog between objects possible, the dialog does not happen by itself. There is a need for some *controller/supervisor* that would coordinate the dialog between objects.

Experience shows that expectations for a system change over time. The more users become familiar with the system, the better they understand the system and their expectations grow; other requirements may be added to the system and the logic that controls the dialog between classes may change. In order to design a flexible system that is resilient to changes, three different aspects of a system need to be taken into consideration. These aspects are:

The communication between users and the system, the control of the logic of the system, and the concepts of the problem domain. Therefore, three different categories of classes, one for each aspect, need to be discovered. The categories are: Boundary, control, and entity classes. Let us take a closer look at each of these categories of classes and present their role in our future system.

Note that the differences between classes that belong to different categories are simply conceptual. Grouping classes into categories helps us to reduce complexity by dividing the problem into smaller and independent parts. During the implementation phase of the system in a particular programming environment, the conceptual differences between classes disappear; they are just classes provided with the right behavior to play well-defined roles in the system.

5.1   Boundary classes

Boundary classes are used to model the interaction between users and the system. As users of the system are modeled as actors, the boundary classes represent the interaction between the actors and the system. Figure 8-10 shows the UML symbols for a boundary class; each of them can be used interchangeably.

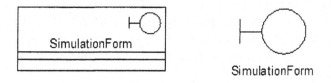

Figure 8-10. UML symbols for a boundary class.

As boundary classes control the interaction between users and the system, actors can communicate only with boundary classes. Boundary classes serve as a shield to separate the internal part of the system from the external events that may affect the system and vice versa. Usually, boundary classes are used to model graphic user interfaces.

There is at the least one boundary class per each actor/use case pair. Figure 8-11 presents an example of using a boundary class, referred to as *SimulationForm*, to control the dialog between the actor, referred to as *user* and the use case *Start Simulation*. The behavior of the class *SimulationForm* should provide all the operations needed to start a simulation, such as entering initial data for soil and plant, selecting a weather station name from a pull down list, etc. Additional behavior such as the one used for validating the input data, can be part of class definition. As the input data are entered using the boundary class, their validation should be part of class's behavior too.

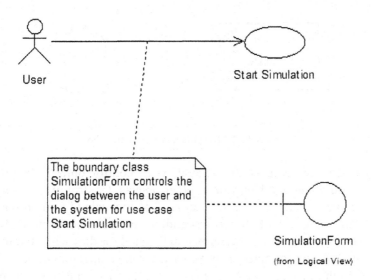

Figure 8-11. One boundary class controls the dialog between an actor and a use case.

It is not necessary at this point of the analysis to go into deep details over how the user interface will be designed and how many items will it contain. These details will be provided later, during the implementation phase. For the moment it is important only to define in a general manner the behavior of the boundary class. As boundary classes are used to model user interfaces, they are platform dependent. As an example, if the implementation environment changes from Windows to UNIX, then the boundary classes will change too. Usually, the lifecycle of a boundary class follows the lifecycle of the corresponding use case. When the use case terminates, there is no more need for the object created from the boundary class.

5.2 Control classes

Control classes are used to model the behavior that is required for the realization of one or more use cases. Thus, a control class should provide the behavior that expresses the realization logic of a use case; therefore, they are use case specific. If the logic of a use case changes, then the behavior of the corresponding control classes should be adjusted accordingly. Figure 8-12 shows the UML symbols for a control class; each of them can be used interchangeably.

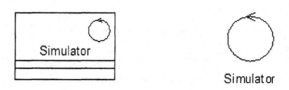

Figure 8-12. UML symbols for a control class.

Objects created from control classes (i.e., control objects) are used to control or coordinate the behavior of other objects. Control objects supervise the flow of events in a use case realization. The lifecycle of a control object is linked to the lifecycle of the corresponding use case. A control object is created when the use case is performed and usually it dies when the use case is terminated. A control object may be used for the realization of one or more use cases. In the case that a use case is complex, many control objects created from different control classes can collaborate to control the use case. The number of control objects that are needed to control a use case is not easily determined. Many factors, such as the designer's experience and the flexibility of the system under construction, can impact the number of control classes that need to be created.

As control classes are closely related to the realization of a use case, they belong to the internal part of the system. Therefore, control classes do not interact with actors; an actor should not communicate directly with a control class. Unlike boundary classes, control classes are platform independent; the same control class can play its role in different computing platforms.

In section 4.1, **Conceptual model for Kraalingen approach,** it is mentioned that classes *Plant*, *Soil*, and *Weather* will provide the necessary data needed in the simulation process. Objects created from these classes will send messages to one another to obtain data located in one object that are required in another object. A controller object is needed to coordinate the interaction amongst objects. Figure 8-13 shows an example of a control class used to control the flow of events in the *Start Simulation* use case.

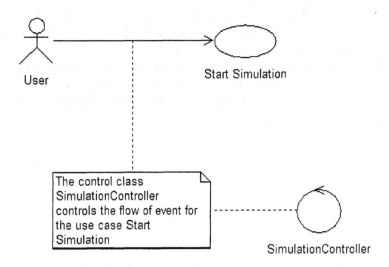

Figure 8-13. SimulationController controls the flow of events for the use case.

5.3   Entity classes

Entity classes are used to model concepts of the system. They are independent of the actors and are usually used to hold and update information about the phenomena under study. Entity objects that are created from entity classes are often persistent objects that need to be stored in a database. Entity classes can be used in many use cases and their behavior can be complex or simple, based on the nature of the problem under study. They can represent real-life objects such as a person, an event, a crop, etc. Figure 8-14 shows UML symbols for entity classes; each of them can be used interchangeably.

Figure 8-14. Icon representations for an entity class

Because entity classes are an internal part of the system, actors cannot communicate with them directly. For similar reasons, boundary classes should not communicate directly with them unless the context of the corresponding use case is simple. Entity classes are provided with behavior that is used to solve the problem. Their main responsibility is to store and manage information in the system.

In the Kraalingen approach, as shown in the conceptual model, classes that hold the data and the behavior needed in the simulation process are *Soil*, *Plant*, and *Weather*. Figure 8-15 shows the entity classes for the *Start Simulation* use case.

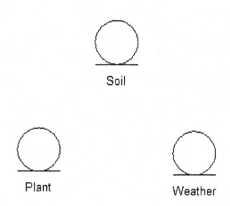

Figure 8-15. Entity classes for the Start Simulation use case.

6.   CLASS DIAGRAM FOR THE KRAALINGEN APPROACH

In the previous sections we discussed issues about how to identify the future classes and their particular role in the system. Now is the time to analyze the relationships between classes in order for the system to provide the required functionality. As we have previously mentioned, the main characteristic of the object-oriented approach is to model concepts of the problem domain using objects and provide objects with data and behavior so that they can play a well-defined role. Objects send messages to one another to carry out functionality. In order for the objects to send messages to each other, they need to have relationships among them. The purpose of the class diagram is to show how objects, created from classes of the system, are interrelated.

In the section on boundary classes, it was mentioned that the class referred to as *SimulationForm* will play the role of a boundary class that is used to model the interaction between users and the system. The main role of this class is to transfer to the system the initial data required for the simulation. Usually, boundary classes are represented as graphical user interfaces. In the case of boundary class *SimulationForm*, the corresponding graphical user interface is shown in Figure 8-16. As shown in this figure, the user interface is divided in two main areas: The area of data input and the area where the results of the simulation are presented. The user will provide the data referred to as the initial conditions for the simulation and the system will provide to the user the results of the simulation. Therefore, objects of class *SimulationForm* should be provided with the appropriate attributes to hold the initial data and the results of the simulation.

Figure 8-16 shows that *SimulationForm* is provided with three buttons: *Simulate*, *Cancel* and *Clear*. The first button allows users to start a simulation process provided that the user has already entered the required initial data. The two other buttons help the user during the data entry process; the user may cancel the session at any time or change all the entry values and replace them with other data.

There is a tendency to create a use case for each of the menu items of the graphical user interface. In our case we would have three use cases: *Start Simulation*, *Cancel Simulation*, and *Clear Simulation*, one use case for each of the functionalities that take place when the corresponding button is used. This modeling practice is not a good one, as it confounds the use cases with menu items. Use cases do not represent and should not represent menu items. By definition, a use case represents the interaction of the user with the system, focusing on what the system can do for the user. In our case, as the only thing that the user can do with the system is to start a simulation, it is appropriate to have only one use case referred to as the *Start Simulation*. The buttons *Cancel* and *Clear* do not provide users with any additional functionality related to the simulation process; the functions they provide are standard functions of any user interface.

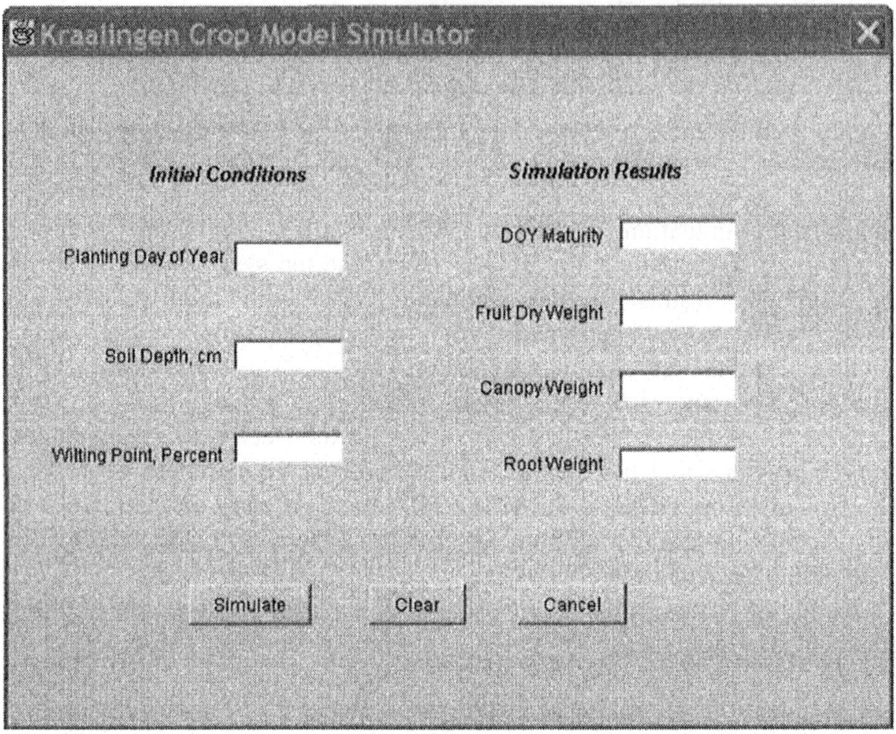

Figure 8-16. Implementation of the boundary class SimulationForm.

In the section on control classes, it is mentioned that a class, referred to as *SimulationController*, will play the role of a control class used to control the behavior of the use case *StartSimulation*. Therefore, the *SimulationController* control class will serve as a coordinator between the boundary class and the system. *SimulationController* will receive the input data from the boundary class and use them accordingly in the simulation process.

The lifecycle of a control object is determined by the boundary class that is related to it. Usually, the boundary object creates an instance of the control class at the beginning of the execution of the use case and the control object "dies" when the use case is terminated.

How does a boundary object pass the input data to a control object? There are several ways to achieve this task. The best solution is to store the input parameters in a file of type *Properties* that will be used by method *simulate(properties)* of class *SimulationController* as shown in Figure 8-17.

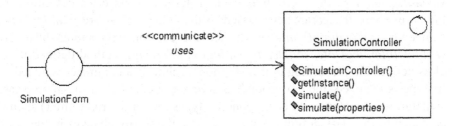

Figure 8-17. A boundary object communicating with a control object.

As shown in this figure, class *SimulationForm* has an association with class *SimulationController* referred to as *uses* of stereotype *communicate*. The diagram says that a *SimulationForm* uses a *SimulationController* to perform a simulation. The association *uses* is unidirectional, meaning that class *SimulationForm* can access data and behavior from class *SimulationController* but not vice versa.

SimulationController is provided with a constructor which is the Java mechanism for creating new instances of a class. As the *SimulationController* is modeled following the *Singleton* pattern, the method *getInstance* returns the unique instance of this class. The definition of class *SimulationController* includes two methods referred to as *simulate* but with different signatures; one does not use any parameters and the other uses a parameter named *properties*.

When method *simulate()* is used, the controller object does not receive any input from the boundary object. In this case, the entity objects used in the simulation should provide all the initial data needed for the simulation process; they will have default values for parameters *plantingDate*, *soilDepth*, and *wiltingPointPercent*. This case is limiting, as it will perform simulations with fixed soil and plant data. The users cannot create scenarios to study the impact of different parameters on crop growth.

When method *simulate(properties)* is used, the boundary object can pass parameters to the control object to be used for populating different entity objects. Parameters are stored in a property file referred to as *properties*. Performing a simulation using initial data is the most common case in the simulation models, as it allows for studying the impact of one or more parameters on crop yield.

Other techniques can be used to pass parameters from a boundary class to the control class. One of them could be to provide the control class with *simulate(list of parameters)* methods that use a different number of parameters. For example, if the parameters passed to the control class are *plantingDate* and *soilDepth*, the control class should have the *simulate(plantingDate, soilDepth)* method defined in its class definition. In the case where the parameters passed to the control class are *plantingDate*, *soilDepth*, and *wiltingPointPercent* then the control class should have the *simulate(plantingDate, soilDepth, wiltingPointPercent)* method defined in its class definition. According to this solution, each time a parameter is added or its type is changed, the corresponding *simulate* method with the appropriate signature (parameters with corresponding types and in the right order) should be added to the class definition of the control class. Any changes in the class definition of the control class implies that the boundary class should be modified accordingly, as the boundary class will pass the simulation parameters to the control class. This solution strongly couples the boundary class to the control class, and as result, the system becomes inflexible. Changes occurring in one class have ramifications in other classes.

Using a container where all the parameters passing to the control class will be stored is a better solution. Such a container in the Java programming environment can be a *Properties* file. Using property files is a simple way in Java to implement a general communication between boundary and control classes. The control class needs only to know that a property file is used to store parameters; the number of parameters passes and their type is irrelevant. This type of communication increases the independence between boundary and control classes. Similarly, an XML file can be used to store the parameters that are passed to the control class.

The control class plays a key role in the simulation process, as it coordinates the messages objects send to each other. Therefore, it is a good programming practice to use the *Singleton* pattern when designing this class. As explained in Chapter 7, **Design Patterns**, the *Singleton* pattern allows for creating only one instance of the class. Method *getInstance* will provide the unique instance of the control class.

In the section on entity classes, it is mentioned that entity classes are used to model concepts of the problem domain. A good start for depicting entity classes is the conceptual diagram for the Kraalingen approach previously developed. According to this diagram, the elements needed for the simulation model are classes *Soil*, *Plant*, and *Weather*. The control class *SimulationController* needs to have access to these objects in order to manage the flow of messages they need to send to each other. Figure 8-18 shows the relationship between controller and entity classes for the Kraalingen approach.

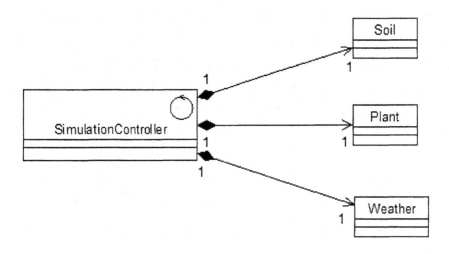

Figure 8-18. Relationship between control and entity classes.

According to Figure 8-18, the relationship between the control and entity classes is modeled as a composition; the controller is conceived as a container that includes entity classes and shields them from the outside view. Thus, the controller manages the lifecycle of the entity objects. Because our problem has only one use case, entity objects are created at the beginning of the use case execution and they "die" when the use case is terminated.

The associations between the control and the entity classes are one-to-one associations. This means that the control class will only create one instance of each of the entity classes. Another important detail presented in Figure 8-18 is the fact that the associations between control and entity classes are unidirectional (i.e., the control object can access data and behavior from entity objects, but the entity objects do not have access to the control object).

Figure 8-19 shows the class diagram for the Kraalingen simulation approach. The diagram says that the *SimulationForm* communicates the input data to the *SimulationController*. The *SimulationController* uses the input data to populate instances of entity classes such as *Plant*, *Soil*, and *Weather*. *Plant* grows in *Soil* and both *Plant* and *Soil* use weather data to calculate their respective processes. When the simulation is terminated, the *SimulationController* will return to the *SimulationForm* the result of the simulation to be displayed to the user.

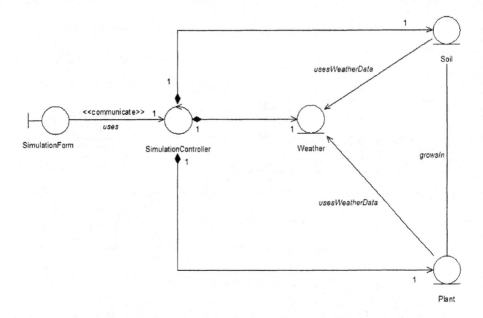

Figure 8-19. Class diagram for the Kraalingen simulation approach.

As shown in Figure 8-19, the *SimulationForm* communicates only with the *SimulationController* as previously mentioned in the **Discover potential classes** section. Thus, the control class incarnates the *Façade* pattern, presented in Chapter 7, **Design Patterns**. Entity objects are protected from the outsiders of the system. Having all the communication with boundary object pass through the control object makes the system independent of the outside environment. Changes that may occur in the behavior of entity objects will not affect the communication with boundary objects.

7.   CRITIQUE OF THE KRAALINGEN CLASS DIAGRAM

In the previous section, we presented a class diagram for the Kraalingen approach that shows the interaction of boundary, control, and entity objects to provide the required functionality. In this section, we will analyze the class diagram in detail to justify each of the steps.

7.1   Communication boundary-control

An important issue that needs particular attention is the communication between boundary and control objects. As previously mentioned, the boundary class is platform dependent and the control class is platform independent. The communication between these two objects should be modeled in such a way that the resulting system is flexible. Changes of requirements, which usually cause changes of boundary class's behavior, should have minimal impact on the system.

The way in which the communication between boundary and control objects is established in the Kraalingen class diagram does not allow for a flexible development. The boundary class has direct connection with the control class; an instance of the control class needs to be created in the boundary object so that the message *simulate* can be sent to this object. In the case that we would like to use another simulation system that provides similar functionalities, the system will not work. The reason is that the boundary object points directly to the *SimulationController* object and will not recognize any other object playing the same role unless the new controller object is referred to as *SimulationController*. Imposing the name of the control class is a considerable limitation. Coupling these two classes directly makes the system less flexible to changes and difficult to reuse.

The solution to this problem is defining one or a set of interfaces that the control class should implement; the behavior of the control class should be expressed using a well-defined set of interfaces. In our simple example, one interface is amply sufficient to describe the services that the control class offers. Figure 8-20 shows an interface defining the services of the control class *SimulationController*. As shown in this figure, the boundary class has an association with the interface *ISimulationController*. Therefore, the boundary object can reach data and behavior from any object created from a class that implements the *ISimulationController* interface. Thus, using an interface instead of a class opens the communication channels between the boundary object and any control object that implements the required interface.

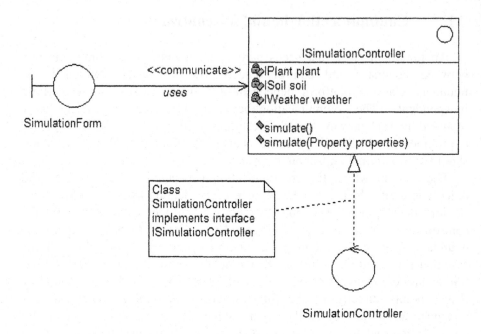

Figure 8-20. An interface defining the services of the control class SimulationController.

The most important advantage of using an interface is that the boundary object does not need to know about the real control object that will receive the simulation parameters. If the simulation system is designed to function as an independent component, any such component can be plugged into a bigger system; the boundary object would communicate with all plugged-in components provided that they implement the required interface. Figure 8-21 shows an example of a boundary class associated with several control classes. Each of the control classes will provide a polymorphic implementation of the behavior defined by the interface *ISimulationController*.

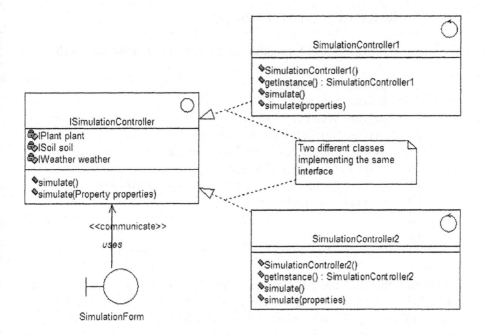

Figure 8-21. One boundary class associated with many control classes.

7.2 Communication control-entity

Another point of communication that needs to be studied carefully is the communication between the control and entity objects. In the class diagram for the Kraalingen approach, the relationships between control and entity classes are modeled as a composition. The control class, *SimulationController*, plays the role of the whole; entity classes, such as *Plant*, *Soil* and *Weather*, play the role of parts. The whole has access to data and behavior of the parts but not vice versa. Objects of the *SimulationController* class have access to entity objects because of the attributes *plant*, *soil*, and *weather* that hold a reference to the corresponding entity objects, as shown in Figure 8-22. This figure presents the Java code for class definition of *SimulationController* modeled as a *Singleton* and defined in the UML diagram shown in Figure 8-18.

Lines 6, 7, and 8 define attributes of types *Plant*, *Soil*, and *Weather* to hold references to the corresponding objects. As we see, the control object is directly connected to the entity objects.

```
1   final class SimulationController {
2
3      private static SimulationController uniqueInstance = null;
4
5      private Plant plant;
6      private Soil soil;
7      private Weather weather;
8      private SimulationController() {}
9      public SimulationController getInstance() {
10        if (uniqueInstance == null)
11           uniqueInstance = new SimulationController();
12        return uniqueInstance;
13     }
14  }
```

Figure 8-22. Java class definition for SimulationController.

The fact that the control object is directly connected to entity objects makes the architecture of the system rigid. When two or more objects are directly linked to each other, none of them can be used separately; the use of one of the objects would require the presence of the others. This architecture is not flexible as it makes it difficult for either the control or entity objects to be reused separately.

The solution to this problem is to avoid linking classes directly; instead, an interface to the class can be used as shown in Figure 8-23. Interface *IPlant* defines the behavior class *Plant* should implement. This behavior includes operations such as *initialize, calculateRate,* and *integrate.* Interface *ISoil* defines the behavior class *Soil* should implement.

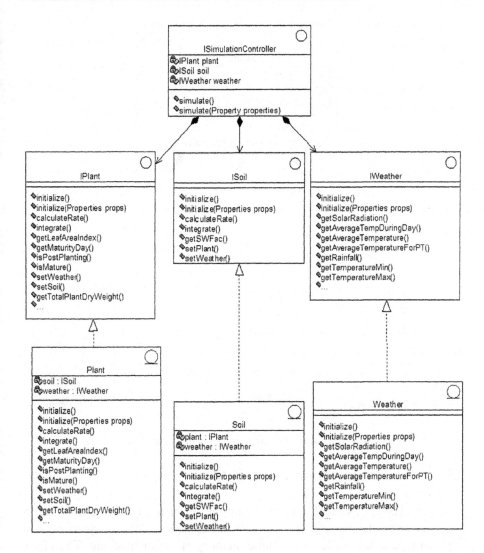

Figure 8-23. The control object communicates with entity objects through interfaces.

Connecting the control class to the interfaces instead of the entity classes opens possibilities to use any other entity class or component that offers similar behavior, provided that they implement the same interface as shown in Figure 8-24.

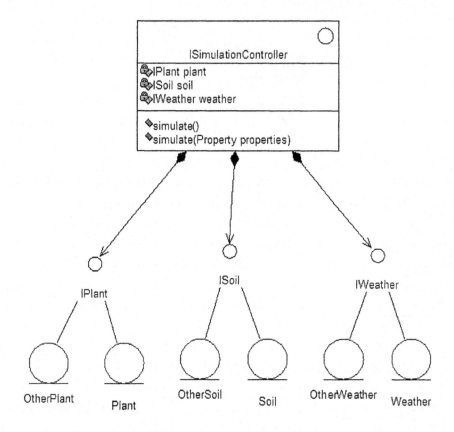

Figure 8-24. Interfaces allow for plugging in other class/components.

7.3   Communication entity-entity

Entity classes represent concepts from the problem domain. They interact with each other to provide the required functionality. Often, objects created from entity classes reside in the same memory space or machine. Therefore, the associations representing the collaborations between concepts are from one entity class directly to another. In this architecture, entity classes are closely coupled to each other. This solution works well in cases that monolithic systems need to be developed; all objects will reside in the same area space and there is no need to substitute the behavior of an object with some other one. Changes to the system can be done only by developers that are familiar with system's architecture. Therefore, these systems are usually not flexible and provide little reuse of functionalities incorporated in the system. There are cases where this architecture is acceptable and provides good results.

There are other cases that although entity objects belong to the same domain, linking them directly to each other may create problems related to the reuse and extendibility of the future system. It would be desirable to have flexible systems that allow for replacing one class or component with some other ones that provide similar functionalities. As an example, some other institutions or individuals may have developed a soil or a plant class/component that we would like to use in our system. In this way knowledge can be transferred easier, collaboration among scientific institutions and/or individuals will be better, and the time for developing a new system will be shorter.

In the case of the Kraalingen approach, the conceptual model shows that entity classes are in association with each other. The collaboration between classes in the conceptual model reflects the fact that plant grows in soil and processes occurring in plant and soil are impacted by the weather conditions. Class *Plant* has an association with class *Soil*; both classes are tightly coupled to each other. The *Plant* class definition has an attribute of type *Soil* that is a reference to an object created from class *Soil*. The situation is similar in the *Soil* class definition; an attribute of type *Plant*, defined in class *Soil*, is a reference to an object of type *Plant*. Objects *Plant* and *Soil* are tightly linked together; none of the objects can be reused separately. The same can be said for the case of each of *Plant* or *Soil* classes in relation with class *Weather*. Objects created from classes *Plant*, *Soil*, and *Weather* form a monolithic system, none of its composing objects can be used separately. Figure 8-25 shows the direct links between the above mentioned classes.

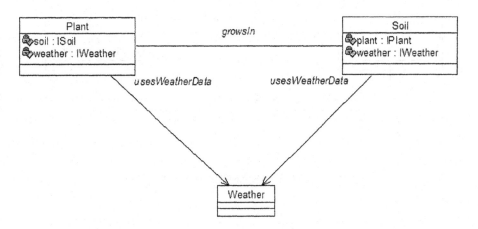

Figure 8-25. A monolithic system where entity classes are directly linked to each other.

In order to make the system flexible and extendable, hardwired associations between classes should be avoided and interfaces be used instead. As we have previously mentioned, the use of an interface is similar to providing a plug that can be used by any class that implements the same interface. The modified class diagram for the Kraalingen approach is shown in Figure 8-26.

As shown in Figure 8-26, the hardwired associations among classes have been replaced with associations between a class and an interface. Therefore, the attribute *soil* in the *Plant* class definition is of type *ISoil*, which is the type of the interface that defines the behavior needed from class *Soil*. Therefore, any class that provides similar behavior to class *Soil* can be used in the system, provided that it implements the interface *ISoil*. Classes implementing the same interface can be used as a substitute to each other. By replacing the association *Plant-Soil* with the association *Plant-ISoil*, we have created the possibility to use any other Soil class that may have a different implementation of the required behavior but it can be used in the system because it implements the interface *ISoil*. The same reasoning could be applied to the associations *Soil-Plant, Plant-Weather, and Soil-Weather*.

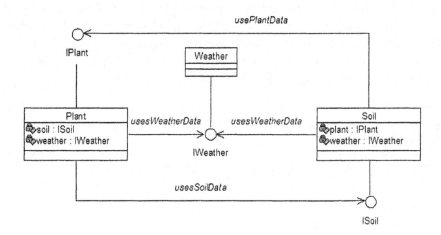

Figure 8-26. Classes have associations with an interface.

8. FINAL CLASS DIAGRAM FOR THE KRAALINGEN APPROACH

Figure 8-27 shows the final class diagram for the Kraalingen approach. For readability reasons, attributes and operations are not presented in the class diagram. Interfaces are used where change or substitutability is expected. For example, the communication point between *SimulationForm* and *SimulationController* is represented by the interface *ISimulationController*. Therefore, any simulation class/component that implements *ISimulationController* can be used. In the same way, the communication point between *SimulationController* and entity classes is represented by three interfaces; *IPlant, ISoil* for communicating with *Plant* and *Soil*, and *IWeather* for communicating with *Weather*.

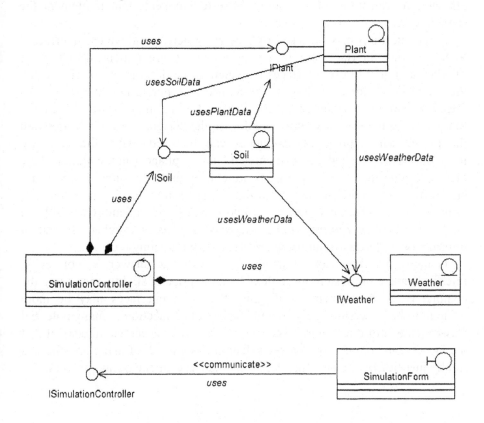

Figure 8-27. Final class diagram for the Kraalingen simulation approach.

9.   THE BENEFITS OF USING INTERFACES

As we have seen in the case of the Kraalingen simulation approach, interfaces are a powerful modeling artifact to design flexible software that can be reused and extended, and is resilient to changes. Interfaces replace the hardwired associations between classes with flexible associations between a class and an interface, opening the way for use to many other potential classes or components that provide similar behavior and implement the same interface. Interfaces make the communication between two classes flexible. When two classes are linked through an association, the link is rigid, as it is a one-to-one link. None of the classes can be used separately. Using interfaces instead of classes transforms the one-to-one link into to a one-to-many link. Therefore, many other classes or components can be used in the system, allowing for substitutability among classes/components that implement the same interface.

Some questions may be asked right away. When should we use interfaces? Should we use interfaces any time that two classes are associated? The issue of when and how to use interfaces is related to the designer's experience and the future needs for extending the system to incorporate new behavior. An interface can be considered as an electrical plug in a house. When the blueprints for the electrical wires are drawn, the designer has in consideration the places where potential appliances can be used. After the electrical installations are finished, one can only add new appliances in the places where there is a plug to be used. Interfaces are conceptually the same. We will use an interface at a point in the system that we see future development, change, or potential for substitutability. We would strongly recommend [CMK99]; the book *Java Design Building Better Apps & Applets* is excellent source of information and ideas about using interfaces as a modeling tool.

Is there any drawback of using interfaces? The answer is, not really. Interfaces make the system more flexible at the price of slightly increasing the complexity of the system. A diagram that uses interfaces may be more difficult to be understood as it contains points of connection to many potential classes instead of one single class. There is a widely accepted agreement that the increase in complexity of a class diagram as a result of using interfaces is rightly justified by the benefits offered by interfaces in the design process.

10. IMPLEMENTATION OF THE KRAALINGEN MODEL IN JAVA

In this section, we will discuss issues related to the implementation of the Kraalingen model in the Java programming environment. The class diagram already created in Section 8, **Final class diagram for the Kraalingen approach**, serves as a starting point for the implementation process. In this diagram, an interface is defined for each of the classes *Weather, Plant,* and *Soil* that formalizes the behavior these classes should provide. The interfaces are referred to as *IPlant, ISoil,* and *IWeather.* Let us take a look at each of the interfaces and examine how they define the behavior of classes implementing the interface.

10.1 Interface IPlant

Interface *IPlant* defines the behavior that needs to be implemented by all classes, providing plant-related behavior that can be used in the simulation. In order to be able to use different classes that provide plant-related behavior, these classes should implement the following interface. Different classes will provide a polymorphic implementation of the same interface. Figure 8-28 shows the Java code for the interface *IPlant*.

```
1   import java.util.Properties;
2   public interface IPlant {
3
4       //used by the simulator controller
5       public void initialize(Properties props);
6       public void initialize();
7       public void calculateRate();
8       public void integrate();
9       public double getLeafAreaIndex();
10      public int getMaturityDay();
11      public boolean isPostPlanting();
12      public boolean isMature();
13      public void setMaturityDay();
14
15      //used to set relationships with other classes
16      public void setWeather(IWeather weather);
17      public void setSoil(ISoil soil);
```

Figure 8-28. Java implementation for interface IPlant (Part 1 of 2).

```
18   //used to provide simulaton results to users
19   public double getFruitDryWeight();
20   public double getRootDryWeight();
21   public double getTotalPlantDryWeight();
22 }
```

Figure 8-28. Java implementation for interface IPlant (Part 2 of 2).

Lines 4, 15, and 19 are comments or stereotypes, defined in Section 11 of Chapter 3 in the first part of the book. They define a specific categorization for the methods that follow the comment or stereotype. Thus, the method *integrate()*, as defined in line 8, belongs to the category referred to as *used by the simulator controller.* Methods defined in lines 20, 21, and 22 belong to the category *used to provide simulation results to users.* Methods defined in the interface *IPlant* are classified in three categories. Classifying methods of an interface in different categories helps one to understand the role of the interface and its behavior. Therefore, we can say that interface *IPlant* is designed to provide three kinds of behaviors: Behavior needed in the simulation process, behavior needed to establish relationships with other classes of the diagram, and behavior to provide simulation results to other objects in the system.

How do we define the behavior of an interface? The behavior of an interface is defined by considering the role classes implementing the interface should play in the system. Interfaces should define one specific role and be defined in its entirety. Two different interfaces should not share common behavior. Designing an interface is a delicate process and needs to be carried out carefully. Interfaces that provide many different behaviors should be avoided. The behavior of such interface should be distributed to other interfaces.

In the case of the class *Plant*, its interface should define all the functions that module *Plant* must provide as defined by [Kra95]. As previously mentioned in this chapter, module *Plant* should be able to perform processes referred to as initialization, rate calculation, and integration. Therefore, the interface *IPlant* includes methods *initialize()*, *calculateRate()*, and *integrate()*, as defined in lines 5 through 8. Note that this interface defines two methods that hold the same name, *initialize*, but with different signatures. The reason for having the same method definition with different signatures is to allow more flexibility in the initialization process of class *Plant*. The method *initialize()*, with no parameters, can be used in cases when users do not provide any initial plant data. In this case, the default data will be used and the *initialize()* method will be activated. The method *initialize(Properties props)* will be used when users want to study the effect of different parameters on plant growth. As an example, the planting date is a parameter that impacts plant growth. This parameter and can be stored in the property file *props* and be used as an entry parameter for the simulation process. In the case that other plant parameters need to be used to study their impact on plant growth, they need to be added to the property file. The following is an example of a property file.

```
plantingDate=121
soilDepth=145
wiltingPointPercent=0.06
```

Method *getLeafAreaIndex()* , defined in line 9, provides information about the leaf area index needed to calculate processes that occur in soil. Methods defined in lines 10 through 13 provide information about the status of the plant; the use of method *isMature()* defined in line 12, will allow other objects to know whether plant has reached the status of maturity. When the plant reaches the status of maturity, the attribute that holds the value of the maturity day needs to be updated. The update process is carried out by the method referred to as *setMaturityDay()*. Method *isPostPlanting()* checks whether the current day in the simulation process is after the planting date. Methods belonging to stereotype *used to set relationships with other classes* allow an object of type *Plant* to have access to objects of type *Soil* and *Weather*. Methods belonging to stereotype *used to display simulation results* represent the behavior that allows an object of type *Plant* to provide results of the simulation to other objects that may request for these data. The object that is most interested to know the results of the simulation is the boundary object referred to as *SimulationForm* object. This object performs the task of communicating with the controller object in order to provide initial values used in the initialization process of plant, soil, and weather and the task of receiving the results of the simulation to display them to the user. In the case that additional plant results are needed, then the necessary methods will be added to the interface definition. As shown in Figure 8-28, interface *IPlant* defines methods that provide dry weight data for the fruit, the root, and the total plant. Any class *Plant* that will be considered for use in our system should implement interface *IPlant*. Figure 8-29 shows the UML diagram representing the association between class *Plant* and its interface *IPlant*.

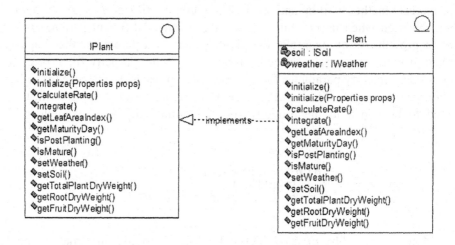

Figure 8-29. Class Plant implements the behavior defined in interface IPlant.

10.2   Interface ISoil

The behavior of interface *ISoil* is defined based on the role classes that implement this interface play in the system. In the case of the class *Soil*, its interface should define all the functions that module *Soil* must provide as defined by [Kra95]. As mentioned in the beginning of this chapter, module *Soil* should be able to read its own initial data, and perform the processes referred to as initialization, rate calculation, and integration. Objects created from class *Soil* should be able to communicate with objects created from class Plant, to obtain from them the plant data required for calculating processes occurring in soil and to establish relationships with other objects in the system. The relationships between objects are already defined in the class diagram presented in Section 7. Figure 8-30 shows the Java code for interface *ISoil*.

As shown in Figure 8-30, there are two stereotypes defined for interface *ISoil*: One referred to as *used by the simulator controller* and the other referred to as *used to set relationships with other classes*. The methods belonging to the first stereotype are used in the simulation process and the methods belonging to the second stereotype allow an object of type *Soil* to set relationships with objects of type *Plant* and *Weather*.

```
1    import java.util.Properties;
2
3    public interface SoilInterface  {
4        public double getSWFac();
5        public void initialize();
6        public void initialize(Properties props);
7        public void calculateRate();
8        public void integrate();
9        public void setWeather(WeatherInterface weather);
10       public void setPlant(PlantInterface plant);
11   }
```

Figure 8-30. Java code for interface ISoil.

Note that lines 5 and 6 define the same method named *initialize()* with different signatures. The method *initialize()*, with no parameters, can be used in cases when users do not provide any initial soil data. In this case, the default soil data will be used and the *initialize()* method will be activated. The method *initialize(Properties props)* will be used when users want to study the impact on plant growth of different soil parameters such as soil depth and wilting point percentage. These parameters will be stored in the property file and are an entry for the simulation process. In the case that other soil parameters need to be considered for the study, their values will be added to the property file. Figure 8-31 shows class *Soil* implementing the *ISoil* interface. Therefore, class *Soil* agrees to provide an implementation for all the methods defined in the interface *ISoil*.

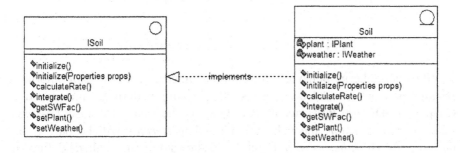

Figure 8-31. Class Soil implements ISoil interface.

10.3   Interface IWeather

The behavior of interface *IWeather* is defined based on the role classes that implement this interface play in the system. The *Weather* class should provide weather data to all other objects in the system that need them. Different sources of weather data can be used; one source could be a text file saved locally in the system and another source could be a network of weather stations that can be accessed on-line. The behavior *IWeather* interface should define is shown in Figure 8-32.

IWeather has a few particularities that make this interface different from other already defined interfaces. All the defined methods are used in the simulation process. There are no methods used to set relationships with other objects as objects of type *Weather*, according to the conceptual diagram, do not have access to other objects of the system. They only provide weather data to all other objects that will request these data.

```
1   import java.util.Iterator;
2   import java.util.Properties;
3
4   public interface IWeather extends Iterator {
5       public double getSolarRadiation();
6       public double getAverageTempDuringDay();
7       public double getAverageTemperature();
8       public double getAverageTemperatureForPT();
9       public double getRainFall();
10      public double getTemperatureMin();
11      public double getTemperatureMax();
12      public double getPAR();
13      public int getDayOfYear();
14      public void initialize(Properties props);
15  }
```

Figure 8-32. Definition of interface IWeather in Java.

IWeather inherits behavior from another interface, the *Iterator* interface. *Iterator* is a pattern and the reasons of using this pattern are introduced in Chapter 7, **Design Patterns** in Part One of the book. Because *IWeather* inherits from *Iterator*, the behavior defined in *Iterator* will be part of the definition of *IWeather* as well. Figure 8-32 shows only the methods defined in interface *IWeather*; methods inherited from *Iterator* are not shown. Any class *Weather* that will be considered for use in the system should implement *IWeather* and *Iterator* interfaces, as shown in Figure 8-33.

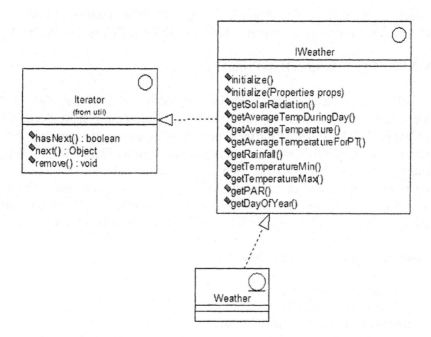

Figure 8-33. Class Weather implements the behavior defined in IWeather and Iterator.

The *Iterator* interface defines the behavior needed to loop over a container of data in order to analyze each of them. The method *hasNext()* is used to assure that the iteration over the data continues as long as there are valid data in the container. The method *next()* gives access to the next data. In the case that there are no available data, *next()* will change the status of *hasNext()* to false and the iteration will stop. The *remove()* method removes an element from the container. This method is not used in our example.

Note that different classes implementing interface *IWeather* will provide different implementations for each of the methods defined in interface *Iterator*. The particular implementation each class will provide will be based on the kind of the container used to hold the data. In the case that the weather data are saved locally in a text file, the data container is a file containing lines and each line contains weather data for a day or other time unit used in the simulation. In the case that weather data are obtained from an on-line weather station, the container is a table and each row of the table contains the weather data for a day or other time unit used in the simulation.

In the next sections, we will show two Weather classes providing different implementations of the same set of *Iterator* and *IWeather* interfaces. The classes are *WeatherDataFromFile* and *WeatherDataFromStation*, as described in Section 6 of Chapter 3, **Interfaces,** in the first part of the book.

10.3.1   Class WeatherDataFromFile

This class implements the behavior defined in *IWeather* interface in the case that the weather data are read from a local file in the system. Section 4.3 of Chapter 7, **The Iterator Pattern,** presents classes that are involved in this pattern and their collaborations. A class referred to as *DailyWeatherData* is designed to hold daily weather data such as solar radiation, minimum and maximum temperature, and rainfall. Note that the simulation unit is the day; an instance of class *DailyWeatherData* contains weather data for a particular day of the year. Figure 8-34 shows the implementation of this class in Java.

```
1  public class DailyWeatherData  {
2
3     private double solarRadiation;
4     private double maxTemperature;
5     private double minTemperature;
6     private double rainFall;
7
8   public DailyWeatherData(String minTemperature,String
      maxTemperature,      String rainFall,String solarRadiation)  {
9    this.solarRadiation=Double.parseDouble(solarRadiation);
10   this.maxTemperature=Double.parseDouble(maxTemperature);
11   this.minTemperature=Double.parseDouble(minTemperature);
12   this.rainFall=Double.parseDouble(rainFall);
13   }
14
15  public double getSolarRadiation() {
16     return solarRadiation;
17  }
```

Figure 8-34. The implementation of class DailyWeatherData in Java (Part 1 of 2).

Lines 3 through 6 define the attributes of the class. Note that these attributes are defined as private; they cannot be accessed outside the class definition by using the attribute name. Lines 8 through 13 define the class constructor, the Java mechanism for creating instances of a class. Lines 15 through 26 define methods to access values attributes hold. Lines 27 through

39 define additional methods for providing other types of daily weather data obtained by manipulating the core daily data.

```
18   public double getTemperatureMax() {
19        return maxTemperature;
20   }
21     public double getTemperatureMin()  {
22        return minTemperature;
23   }
24   public double getRainFall()  {
25      return rainFall;
26   }
27   public double getPAR()  {
28      return 0.50*getSolarRadiation();
29   }
30   public double getAverageTempDuringDay()  {
31      return 0.6*getTemperatureMax()+0.4*getTemperatureMin();
32   }
33     public double getAverageTemperatureForPT() {
34        return
       0.25*getTemperatureMin()+0.75*getTemperatureMax();
35   }
36   public double getAverageTemperature() {
37        return
       0.5*getTemperatureMin()+0.5*getTemperatureMax();
38   }
39 }
```

Figure 8-34. The implementation of class DailyWeatherData in Java (Part 2 of 2).

As shown in Figure 8-34, once an instance of the class *DailyWeatherData* is created, individual data can be obtained using its accessor methods. As an example, the method *getTemperatureMin()* sent to this object will return the value of the minimum temperature. Note that the definition of this class does not depend on the container used for holding the weather data.

Figure 8-35 shows the implementation in Java for class *WeatherDataFromFile*. The role of this class in the simulation process is to provide weather data stored locally in a text file. Lines 1 and 2 show the libraries that need to be imported for the class definition. These are the input-output and utility libraries. The utility library contains the *Properties* utility, used in line 12. Line 3 defines the class *WeatherDataFromFile*, stating that this class implements the interface *IWeather*. Because *IWeather* inherits from interface *Iterator*, class *WeatherDataFromFile* needs to implement behavior defined in both interfaces; in addition to the behavior defined in *IWeather* interface, this class should also implement the behavior defined in the *Iterator* interface. *Iterator* is an interface defined in the *java.util.Iterator* library that is provided by the Java development environment. The same Java environment provides *java.io.* * library.

```
1     import java.io.*;
2     import java.util.*;
3     public class WeatherDataFromFile implements IWeather  {
4
5        private int dayOfYear;
6        private BufferedReader br=null;
7        private DailyWeatherData currentDay=null;
8        public WeatherDataFromFile(){}
9
10       public void initialize(Properties props) {
11          try {
12             String fileName = props.getProperty("weatherFile");
13             FileReader fileReader = new FileReader(fileName);
14             br = new BufferedReader(fileReader);
15             setDayOfYear(0);
16             }
17          catch (FileNotFoundException e)
18             {
19             System.out.println("Weather file not found; the system will shut down");
20             System.exit(1); // Implementation of the precondition
21             }
22          catch (IOException e){System.out.println("IO Exception");}
23          }
24       public boolean hasNext() {
25          try {
26             return br.ready();
27             }
28          catch (IOException e){return false;}
29          }
30       public Object next() {
```

Figure 8-35. Definition of class WeatherDataFromFile in Java (Part 1 of 4).

Lines 5 through 7 define the attributes of class *WeatherDataFromFile*. Note that line 7 defines an attribute of type *DailyWeatherData*; its implementation details were presented in Figure 8-34. Line 8 defines the class constructor; in this case, it is a default constructor.

Lines 10 through 23 define the body of the method *initialize(Properties props)*. The scope of this method is to prepare the environment for obtaining the weather data. An instance of the class *FileReader* is created using the current value of the attribute *filename* read from the property file. The property file is made available to object *Weather* by the simulator controller. In the case that the weather file is not found, the system halts the execution and displays an error message, as shown in lines 19 and 20. These lines (19 and 20) show the implementation of the precondition for the use case as mentioned in Section 2.4, **Preconditions,** in this chapter. Then, an instance of class *BufferReader* is created using the file reader already obtained and the day of the year is set to zero. The input-output library imported in line 1 provides the functionality required to read data from a text file. Lines 24 through 29 show the implementation of method *hasNext()* defined in the interface *Iterator*. This method will return the result **true** when data are available and **false** otherwise. Lines 30 through 44 define the method *next()* that provides the next set of available weather data. Line 32 reads a line from the data container. Line 33 divides the entire line into tokens. Tokens are created by considering the values separated by comma.

```
31      try {
32      String line = (String)br.readLine();
33      StringTokenizer tokens = new StringTokenizer(line,",");
34      if (line.length()>0)  {
35         String [] dailyData = new String [tokens.countTokens()];
36         dailyData[0] = tokens.nextToken();
37         dailyData[1] = tokens.nextToken();
38         dailyData[2] = tokens.nextToken();
39         dailyData[3] = tokens.nextToken();
40         dailyData[4] = tokens.nextToken();
41         currentDay=new DailyWeatherData(dailyData[1],
        dailyData[2], dailyData[3], dailyData[4]);
42      }
43      }
44      catch (IOException e){System.out.println("Error reading data");}
```

Figure 8-35. Definition of class WeatherDataFromFile in Java (Part 2 of 4).

Line 34 makes sure that the string read in line 32 contains real data. Line 35 creates an array with size the number of tokens created in line 33. Lines 36 through 40 assign to an element of an array a token that represents a value (rainfall, for example) from the daily data set. Line 41 creates an instance of the class *DailyWeatherData* with the tokens obtained previously. The values of tokens will be assigned to the attributes of class *DailyWeatherData*. Line 44 shows an exception that may occur while reading the data from the text file. It is a good programming practice to provide users with the right information when an exception occurs during the execution of the program. The user is informed about the cause of the exception and then, can decide what decision to make next.

```
45      setDayOfYear(getDayOfYear()+1);
46      return currentDay;
47      }
48      public void remove(){}
49      public void setDayOfYear(int dayOfYear) {
50          this.dayOfYear=dayOfYear;
51      }
52      public int getDayOfYear() {
53          return dayOfYear;
54      }
55      public DailyWeatherData getDailyData() {
56          return currentDay;
57      }
58      public double getSolarRadiation() {
59          return currentDay.getSolarRadiation();
60      }
```

Figure 8-35. Definition of class WeatherDataFromFile in Java (Part 3 of 4).

Line 45 increases by one the number of days since the beginning of the simulation. Line 46 returns an instance of the class *DailyWeatherData* populated with current weather data read from the file. The order of the weather data in the file is the following: Solar radiation, temperature max, temperature min, and rainfall.

Line 48 is the definition of the method *remove()*; this method does not have any body, as it is not used. We are obliged to provide an empty implementation for this method as it is part of the interface *Iterator*. Class *WeatherDataFromFile* implements interface *Iterator* and therefore, an implementation for each of the methods of the interface is needed to be part of the class definition. An empty implementation for a method means that the method does not provide any functionality. Lines 49 through 51 define the method that can change the value of attribute *dayOfYear*. This method uses the parameter *dayOfYear* to substitute the existing value of the attribute. Lines 52 through 54 define a method that returns the value of the attribute *dayOfYear*. Lines 55 through 57 define the method *getDailyData()* that returns an instance of class *DailyData* already populated with weather data for a specific day.

```
61  public double getTemperatureMax() {
62      return currentDay.getTemperatureMax();
63  }
64  public double getTemperatureMin() {
65      return currentDay.getTemperatureMin();
66  }
67  public double getRainFall() {
68      return currentDay.getRainFall();
69  }
70  public double getPAR() {
71      return currentDay.getPAR();
72  }
73  public double getAverageTempDuringDay() {
74      return currentDay.getAverageTempDuringDay();
75  }
76  public double getAverageTemperatureForPT() {
77      return currentDay.getAverageTemperatureForPT();
78  }
79  public double getAverageTemperature() {
80      return currentDay.getAverageTemperature();
81  }
82  }
```

Figure 8-35. Definition of class WeatherDataFromFile in Java (Part 4 of 4).

Lines 58 through 82 define methods that allow other objects to access specific weather data stored in the current instance of *DailyWeatherData*. Note that these methods use the **Delegation pattern**, defined in the first part of the book in Section 2.1 of Chapter 7. As class *WeatherDatafromFile* implements interface *IWeather*, it should provide an implementation of all of the methods defined in the interface. Data such as rainfall or solar radiation are stored in an instance of class *DailyWeatherData*. *WeatherDataFromFile* does not have access to these individual data, but it has access to the instance of *DailyWeatherData* that holds them and therefore it delegates the method call to this instance. The format of the file holding the weather data is shown in Figure 8-36. The first column shows that it is the first day of the year 1987. The next columns hold the data for temperature minimum, temperature maximum, solar radiation, and rainfall. In our implementation of the Kraalingen approach we have used a slightly different approach to obtain the day of the year. Instead of extracting it from the *DayOfYear* value, we have defined a new attribute named *dayOfYear* that initially is set to zero and is increased by one at each step of the simulation.

```
DayOfYear   Tmin   Tmax   Radiation Rainfall
---------   ----   ----   --------- --------
87001,      5.1,   20,    4.4,      23.9
87002,      10.8,  13.3,  1.1,      0
87003,      12.1,  14.4,  1.1,      0
87004,      3.6,   18.3,  6.1,      14.7
87005,      12.8,  17.2,  5.6,      0.8
87006,      12.4,  21.1,  5,        0
87007,      11.1,  21.7,  7.2,      0
87008,      12,    21.7,  8.3,      0
87009,      6.1,   20,    8.3,      0
87010,      3.5,   23.3,  11.1,     3.8
```

Figure 8-36. The format of the weather data file.

10.3.2　Class WeatherDataFromStation

This class implements the behavior defined in *IWeather* interface in the case that the weather data are obtained from an on-line weather station. In this case, additional information needs to be provided such as the name or the identification number of the weather station and the starting and the ending date for the time interval of the simulation. These input data are combined in an SQL statement to extract the corresponding records from the database. Figure 8-37 shows the Java implementation of class *WeatherDataFromStation*.

```
1    import java.net.*;
2    import java.util.*;
3    import java.io.*;
4
5  public class WeatherDataFromStation implements IWeather  {
6  private int dayOfYear;
7  private DailyWeatherData currentDay;
8  private BufferedReader buffReader = null;
9  private String startingDate, endingDate, stationNumber;
10 private URL url=null;
11
12 public WeatherDataFromStation(){}
13
14 public void initialize(Properties props) {
15       try {
16          startingDate = props.getProperty ("startingDate");
17          endingDate = props.getProperty("endingDate");
18          stationNumber = props.getProperty ("stationNumber");
19          url=new URL("http://fawn.ifas.ufl.edu/scripts/fawndataserver.asp?
             sql=select%20AirTempMin,AirTemMax,Rainfall,TotalRad
             %20from%20dailysummary%20
20             where%20id="+stationNumber+"%20and%20datetime>="'
21             +startingDate+'"%20and%20datetime<"'
22             +endingDate+'"');
23       }
24    catch (MalformedURLException me) {
25
26             System.out.println("Cannot connect to the weather station");
27             System.exit(1);
28             }
```

Figure 8-37. Definition of class WeatherDataFromStation in Java (Part 1 of 5).

Lines 1 through 3 import the libraries needed for implementing the behavior of the class. Notice that in addition to the libraries needed in the definition of the class *WeatherDataFromFile*, there is another one referred to as *java.net.** that provides behavior to communicate with the Internet. Line 5 defines the class *WeatherDataFromStation* that implements behavior defined in interface *IWeather*. Lines 6 through 10 define the attributes of the class. Line 12 defines the constructor of the class. Line 14 defines the method *initialize(Properties props)*. The property file *props* holds the parameters needed to identify the weather station. The parameters are the weather station number, the starting and ending date for the time interval used in the simulation. Lines 16 through 18 read the above mentioned parameters from the property file. Lines 19 through 23 create an instance of type *URL* using a parameter that provides information about the address of the server and the SQL1 statement to be executed by the server providing the data. Lines 26 and 27 show that in the case the system cannot connect to the weather station, an error message is displayed and the execution halts.

Lines 29 through 35 read the first line of data from the server and set to 0 the attribute *dayOfYear*. The address of the server is: http://fawn.ifas.ufl.edu /scripts/fawndataserver.asp and the SQL statement is:

SELECT   AirTempMin,AirTempMax,Rainfall,TotalRad,ET   FROM dailysummary WHERE id=stationNumber
AND datetime>= startingDate AND datetime<endingDate .

Note that some additional characters are needed to fill the empty spaces in the SQL statement in Java language. The rest of the code is shown as follows.

```
29 try {
30      buffReader = new BufferedReader(new
                 InputStreamReader(url.openStream()));
31    String firstLine = (String)buffReader.readLine(); //eliminates the titles
32 }
33 catch (IOException e){e.printStackTrace();}
34    setDayOfYear(0);
35 }
36 public boolean hasNext() {
37    try {
38      return buffReader.ready();
39      }
40    catch (IOException e) {
41      return false;
42      }
43 }
44 public Object next()  {
45    try {
46      String line = (String)buffReader.readLine();
47      StringTokenizer tokens = new StringTokenizer(line,",");
48      if (line.length()>0) {
49        String [] dailyData = new String [tokens.countTokens()];
50        dailyData[0] = tokens.nextToken();
51        dailyData[1] = tokens.nextToken();
```

Figure 8-37. Definition of class WeatherDataFromStation in Java (Part 2 of 5)

```
52        dailyData[2] = tokens.nextToken();
53        dailyData[3] = tokens.nextToken();
54        currentDay=new DailyWeatherData( dailyData[0],
                     dailyData[1],dailyData[2],dailyData[3]);
55     }
56  }
57  catch (IOException e){}
58  setDayOfYear(getDayOfYear()+1);
59  return currentDay;
60  }
```

Figure 8-37. Definition of class WeatherDataFromStation in Java (Part 3 of 5)

Note that the Java code for classes *WeatherDataFromFile* and *WeatherDataFromStation* are similar. The difference is the part of the code that identifies the source of the data. In the case of class *WeatherDataFromFile*, special code is needed to read the data from the file whereas in the case of class *WeatherDataFromStation*, special code is needed to establish connection with the weather station. In both cases, method *initialize(Properties props)* is used to connect with the data source.

```
61  public void setDayOfYear(int dayOfYear) {
62      this.dayOfYear=dayOfYear;
63  }
64  public int getDayOfYear() {
65      return dayOfYear;
66  }
67   public double getAverageTemperature() {
68       return currentDay.getAverageTemperature();
69   }
70  public void remove(){}
71  public double getSolarRadiation() {
72      return currentDay.getSolarRadiation();
73  }
74  public double getAverageTempDuringDay() {
75      return currentDay.getAverageTemperature();
76  }
77  public double getAverageTemperatureForPT() {
78      return currentDay.getAverageTemperatureForPT();
79  }
```

Figure 8-37. Definition of class WeatherDataFromStation in Java (Part 4 of 5)

```
80      public double getRainFall() {
81         return currentDay.getRainFall();
82      }
83      public double getTemperatureMin() {
84         return currentDay.getTemperatureMin();
85      }
86      public double getTemperatureMax() {
87         return currentDay.getTemperatureMax();
88      }
89      public double getPAR() {
90         return currentDay.getPAR();
91      }
92      public double getPotentialET() {
93         return currentDay.getPotentialET();
94      }
95      }
96
```

Figure 8-37. Definition of class WeatherDataFromStation in Java (Part 5 of 5).

The rest of the class definition in Figure 8-37 is similar to the class definition of *WeatherDataFromFile*. Here again we have used the **Delegation pattern** to delegate a method call from class *WeatherDataFromStation* to class *DailyWeatherData*. As an example, lines 92 through 94 define the method *getPotentialET()*; the method call is delegated to object *currentDay* of type *DailyWeatherData*.

10.4   Interface ISimulationController

The behavior of the interface *ISimulationController* is defined based on the role class *SimulationController* must play in the system. This is a control class as defined in Section 5.2, **Control Classes,** in this chapter. In this section, we mentioned that the role of control classes is to coordinate the interaction of different objects used in a use case realization. Therefore, objects created from the *SimulationController* class should have access to other objects used in the use case *Start Simulation,* to send them the right message at the right time. Objects used in this use case are of type *Plant*, *Soil*, and *Weather*, (i.e., objects created from classes *Plant, Soil* and *Weather*).

Furthermore, as class *SimulationController* controls the dialog with the user interface (or the boundary class), we will model it to follow the **Façade Pattern** as defined in Section 4.4 of Chapter 7. According to this pattern, an object of type *SimulationController* will be the unique point of communication between the user interface or the boundary object and all other objects involved in the *Start Simulation* use case. The interface *ISimulator* should define the operations needed to communicate with the user interface or the boundary object. Figure 8-38 shows the Java implementation of the interface *ISimulator*.

```
1    import java.util.Properties;
2
3    public interface ISimulationController {
4        public void simulate(Properties props);
5        public Properties getSimulationResults();
6    }
```

Figure 8-38. Definition of interface ISimulationController in Java.

As shown in Figure 8-38, the interface *ISimulationController* defines the behavior that needs to be implemented by all classes that are candidates to play the role of the controller in the simulation process. Line 4 defines the method *simulate(Properties props)* that passes to a *SimulationController* object a parameter of type *Properties* that holds the initial values to be used for instantiating objects of classes *Plant*, *Soil*, and *Weather*. Line 5 defines the method *getSimulationResults()* that will be used by the boundary object to obtain the result of the simulation. Therefore, *SimulationController* will store the results of the simulation in an object of type *Properties*.

It is important to note that the provided solution establishes a communication bridge only between the boundary and controller objects. The boundary object does not have access to entity objects such as *plant*, *soil*, or *weather*. The boundary object can communicate to all controller objects, provided they implement the required interface as defined in Figure 8-38. Figure 8-39 shows the implementation of class *SimulationController* in Java.

In this figure, line 1 shows that class Properties is imported from the Java library system. Line 3 defines class *SimulationController* using the **Singleton** pattern and implementing the interface *ISimulationController*. Line 4 defines an attribute of type *SimulationController* referred to as *uniqueInstance*, as this attribute will hold the unique instance of the class. Lines 6, 7, and 8 define attributes of types *IPlant*, *ISoil*, and *IWeather* that respectively reference objects *plant*, *soil*, and *weather*. Line 9 defines an attribute of type *Properties* that is used to hold the results of the simulation. Lines 11 through 24 define a constructor for the class. As the class is modeled using the **Singleton** pattern, the constructor is **private**, meaning that no other object in the system can call this method. Therefore, only one instance of the class can be created. Lines 12, 13, and 14 create instances for objects *plant*, *soil*, and *weather*. The corresponding classes *Plant*, *Soil* and *WeatherDataFromFile*, implement the required interfaces: *IPlant*, *ISoil*, and *IWeather*.

```
1    import java.util.Properties;
2
3    finalpublic        class        SimulationController        implements
     ISimulationController{
4        private static SimulationController uniqueInstance=null;
5
6        private IPlant plant;
7        private ISoil soil;
8        private IWeather weather;
9        private Properties props;
10
11       private SimulationController() {
12           plant = new Plant();
13           soil = new Soil();
14           weather = new WeatherDataFromFile();
15
16           //Establish relation Soil-Plant
17           plant.setSoil(soil);
18           soil.setPlant(plant);
19
20           //Establish relation Plant-Weather
21           plant.setWeather(weather);
22           //Establish relation Soil-Weather
23           soil.setWeather(weather);
24       }
25
26       public static SimulationController getInstance() {
27         if (uniqueInstance == null)
28         uniqueInstance = new  SimulationController();
29
30         return uniqueInstance;
31       }
32
33       public void simulate(Properties props) {
34           // Initializations
```

Figure 8-39. Implementation of class SimulationController in Java (part 1 of 3).

```
35        weather.initialize(props);
36        soil.initialize(props);
37        plant.initialize(props);
38
39        while (weather.hasNext()) {
40           weather.next();
41
42           soil.calculateRate();
43           if (plant.isPostPlanting())
44               plant.calculateRate();
45
46           soil.integrate();
47           if (plant.isPostPlanting())
48               plant.integrate();
49
```

Figure 8-39. Implementation of class SimulationController in Java (part 2 of 3).

Lines 16 through 24 establish the relationships between created objects as required by the conceptual diagram presented in section 4.1 **Conceptual Model for the Kraalingen Approach**. Lines 26 through 31 define method *getInstance()* that returns the unique instance of the class. This method implements the **lazy initialization** principle, which requires instantiating an object only when it is needed. Lines 33 through 56 show the implementation code for method *simulate(Properties props)*. Lines 34 through 37 initialize the three entity objects involved in the simulation process. Note that some of the initial values for populating each of the entity objects are stored in the *props* file. Each of the objects will read the appropriate data from the *props* file. Lines 39 through 55 show the iteration over the weather data. At each step of the simulation (i.e., every day), the processes of rate calculation and integration take place. Lines 50 through 54 implement the repetition condition; if plant has not yet reached the status of maturity, then the simulation will continue. The simulation will terminate if the plant reaches the status of maturity and the final results will be stored in the *props* file.

```
50            if (plant.isMature()) {
51                plant.setMaturityDay();
52                saveFinalResults();
53                return;
54            }
55        }
56    }
57    public Properties getPropertyFile() {
58        return props;
59    }
60
61    private void saveFinalResults()  {
62
63            props = new Properties();
64            props.put("totalPlantDryWeight",
                        new Double( plant.getTotalPlantDryWeight()));
65            props.put("rootDryWeight",
                        new Double(plant.getRootDryWeight()));
66            props.put("fruitDryWeight",
                        new Double(plant.getFruitDryWeight()));
67            props.put("maturityDay",new Integer (plant.getMaturityDay()));
68        }
69 }
```

Figure 8-39. Implementation of class SimulationController in Java (Part 3 of 3).

Lines 57 through 59 define the method *getPropertyFile()* that returns the property file where the results of the simulation are saved. Lines 61 through 69 show the implementation of the method *saveFinalResults()*. Plant data such as total plant dry weight, root dry weight, fruit dry weight, and the day of the year the maturity status is reached are saved in the props file. Therefore, these final plant results can be used by the boundary object to display them to the user.

11.   PACKAGING THE APPLICATION

Before starting to write code, it is important to create a flexible and logical structure for storing files. Files could be organized in packages: A package for component. The entire application is stored in the package referred to as *Kraalingen*, as shown in Figure 8-40. Within this package, five other packages are defined named *Interfaces*, *Plant*, *Simulator*, *Soil*, and *Weather*.

Each of the packages contains a file of type JAR that is the compressed code for the class/component. Thus, the package *Interfaces* contains the file *Interfaces.jar*, the package *Plant* contains the file *Plant.jar*, and so on.

Figure 8-40. Package structure for the Kraalingen Application.

Documents pertaining to the entire system can be stored in the main directory referred to as *Kraalingen*. Thus, the UML diagrams such as class, sequence, and collaboration diagrams can be stored in package *Kraalingen*.

Chapter 9

THE PLUG AND PLAY ARCHITECTURE

1.   DEFINITION

Software engineers are facing increasing pressure from clients to provide architectural solutions that can be built with today's requirements and be flexible enough to meet future needs. It is common practice in the domain of software development to see customers constantly modifying and expanding the requirements of the future system. Faced with the reality of everchanging requirements, it is desirable to use an architecture that allows for easy modifications of existing functionalities and easy adoption of new ones.

An architecture that is designed to minimize the impact of future changes is the "plug and play" architecture. According to the definition provided by Webopedia (http://webopedia.com), the plug and play architecture refers to *the ability of a computer system to automatically configure expansion boards and other devices*. Originally, the plug and play architecture was used by the hardware industry. The idea of installing a new device that will configure itself to work in harmony with other parts of an existing system is very promising. The "plug and play" architecture eliminated the need to adjust switches, jumpers, and other configuration elements in a hardware system. It brought general relief to the frustration caused by the large number of the problems encountered during the process of installing a new piece of hardware. The success of the plug and play approach in the hardware industry created a fertile environment in the software engineering environments to build software systems applying the same techniques.

The key to building software systems that provide this high level of flexibility is to use a component-based design. Components are units that can be developed independently and even from third parties. They can be organized to dialog with each other and to provide the functionalities required by a software system. Components provide their services through a well-defined set of interfaces. It is important to note that defining a well-thought-out set of interfaces is crucial to a component's use and reuse. If interfaces are not fully encapsulating, it will be difficult to tune or enhance implementations, as poor encapsulation will hinder reuse [Lak96]. When the component-based development reached its level of maturity, the plug and play technology was at programmers reach. Plug-ins have been used widely since the introduction of Netscape's Navigator Web browser and one of the most successful examples of plug-in architecture was Apple's QuickTime [Szy99].

2.   IMPLEMENTATION

The implementation of the plug and play architecture is closely related to the use design patterns, specifically to creational and behavioral patterns. A plug and play architecture for the Kraalingen approach should allow the flexibility to replace the basic class/components (i.e., *Soil*, *Plant*, and *Weather*) with other components that provide similar behavior and implement the same set of interfaces as the exiting class/components [Pap05]. In Section 7.2, **Communication control-entity,** in Chapter 8, we mentioned that the control object communicates with entity objects through interfaces. Therefore, any class implementing the interface could be used in the system. It is desirable to have a mechanism that allows for substituting a class/component with another similar one that does not require changes of code. Ideally, such a mechanism would make the choice of the class/component to be used at run time. The problem of selecting among many potential class/components is solved by the **Strategy** pattern, presented in Section 5.2 of Chapter 7. The strategy pattern used for implementing the plug and play architecture is shown in Figure 9-1.

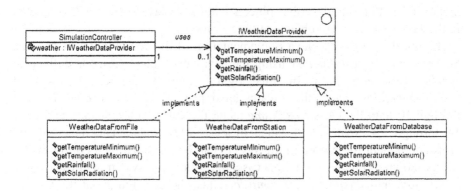

Figure 9-1. Strategy pattern used for implementing a plug and play architecture.

Selecting for use a class at run time requires a mechanism for dynamically creating instances of used classes. The *AbstractFactory* pattern will be used for this purpose. The Java implementation of the patterns used is closely related to the concept of Reflection, discussed in the next section.

3. REFLECTION

Complex and dynamic systems allow for the fact that the environment in which they run may change constantly. In an object-oriented environment, classes are loaded dynamically, binding is done dynamically, and object instances are created dynamically when they are needed. Therefore, there is a need to collect information about the object that is dynamically created. Reflection provides the answer to the above problem. Both Java and .NET technologies provide ample support for reflection. The .NET framework uses reflection to inspect the content of assemblies [Pro02] and Java uses it to collect internal information about classes or components (http://javasoft.com). As Java is the implementation language for our application, only details for Java's Reflection API are provided.

In Java, the Reflection API has a two fold purpose. First, it provides a mechanism to fetch data about a class/component and second, a means for extracting objects composing the class/component. Using reflection, it is possible to obtain internal information about the class/component such as its superclass, the interfaces the class implements, the methods, their signatures, and the returning object. The behavior for fetching data about a class is provided in a class referred to as *Class*. *Class* is the universal type for the meta information that describes objects within the Java system. Class loaders in the Java system return objects of type *Class*.

Figure 9-2 shows the Java implementation of the combination of patterns *Strategy* and *AbstractFactory* using the behavior of class *Class* as defined in Java's Reflection API. In this figure, line 3 defines the method *newInstance(className)* that creates an object of type *className*. Lines 5 through 10 load the class *className* using the current class loader, and lines 11 through 24 create an instance of class *className*. The name of the class/component to be used in the system can be provided by a configuration file as shown in Figure 9-3. In this case, the simulation system will use classes *Plant*, *Soil*, and *WeatherDataFromFile*. Thus, the decision about the type of objects to be created can be made during execution by a component-controller, not in the source code. In order to plugin another class, for example, *WeatherDataFromDatabase*, only the content of the configuration file needs to be updated with the appropriate class name. Thus, no changes to the code are required. The component-controller, in our case *SimulationController*, instantiates the right components at run time. Using the plug and play architecture, the user has the choice to activate different classes that provide similar behavior but implement the same interface.

```
1 public class ObjectFactory {
2
3    public static Object newInstance(String className){
4    Class cls = null;
5    try {
6        cls = Class.forName(className);// create the class
7    }
8    catch (ClassNotFoundException cnfe){
9        System.out.println("can't find class named: " +className);
10   }
11   Object newObject = null;
12   if (cls != null) {
13       try {
14           newObject = cls.newInstance(); // Create the instance
15       }
16       catch (InstantiationException ie) {
17           System.out.println("can't instantiate class named: " +className);
18       }
19       catch (IllegalAccessException iae) {
20           System.out.println("can't access class named: " + className);
```

Figure 9-2. The implementation in Java of Strategy and AbstractFactory patterns (Part 1 of 2).

```
21      }
22    }
23    return newObject;
24    }
25 }
```

Figure 9-2. The implementation in Java of Strategy and AbstractFactory patterns (Part 2 of 2).

```
plant=Plant
soil=Soil
weather=WeatherDataFromFile
```

Figure 9-3. Example of configuration file.

4.   THE PLUG AND PLAY SIMULATORCONTROLLER

As the process of instantiating the class/components involved in the simulation process changed (i.e., the selection of class/components to be used will be done at run time), the controller class, defined in section 10.4 of Chapter 8, must be modified accordingly. Figure 9-4 shows the Java implementation of the modified *SimulationController* class.

```
1     import java.util.Properties;
2     import java.util.ResourceBundle;
3
4     final public class SimulationController  {
5
6        private static SimulationController uniqueInstance=null;
7        private IPlant plant;
8        private ISoil soil;
9        private IWeatherDataProvider weather;
10       private Properties props;
```

Figure 9-4. Java implementation of the plug and play SimulatorController (Part 1 of 4).

```
11        private SimulationController() {
12            ResourceBundle classBundle =
                    ResourceBundle.getBundle("ClassNames");
13            weather = (IWeather)ObjectFactory.newInstance (
                    classBundle.getString("weather"));

14            plant = (IPlant)ObjectFactory.newInstance(
                    classBundle.getString("plant"));
15            soil = (ISoil)ObjectFactory.newInstance(
                    classBundle.getString("soil"));
16            //Establish relation Soil-Plant
17            plant.setSoil(soil);
18            soil.setPlant(plant);
19            //Estalish relation Plant-Weather
20            plant.setWeather(weather);
21            //Establish relation Soil-Weather
22            soil.setWeather(weather);
23        }
24
25        public static SimulationController getInstance() {
26            if (uniqueInstance == null)
27                uniqueInstance = new  SimulationController();
```

Figure 9-4. Java implementation of the plug and play SimulatorController (Part 2 of 4).

Line 2 imports a library that makes available the functionalities of class *ResourceBoundle.* The functionalities provided by this class are used to create classes from names read from the configuration file, as shown in lines 13 through 16. Lines 7 through 9 show that *SimulationController* has access to interfaces *IPlant, ISoil,* and *IWeatherDataProvider.* Line 13 defines a resource bundle, a Java artifact for implementing a configuration file that will hold the names of the classes that *SimulationController* needs to create. Figure 9-2, in the previous section, shows an implementation example of the configuration file. This file can be edited using any text editor. In this case, the *SimulationController* needs to create classes *Plant, Soil,* and *WeatherDataFromFile.* Line 13 obtains access to the configuration file referred to as *ClassNames.* Lines 14 through 16 create objects of type *Weather, Plant,* and *Soil.* As the process of creating objects using reflection is not a very straightforward one, let us take a closer look at this process. First, the name of the class is obtained as a string by sending to *classBundle* the message *getString("className")*. Thus, *classBundle.getString ("weather")* will return the name of class *Weather*. Second, *(IWeather) ObjectFactory.newInstance(classBundle.getString("weather"))* will create an object of type *IWeather*. Note that ObjectFactory returns an object that is casted to become an instance of type IWeather. The same process is used to obtain objects of types *Plant* and *Soil*. Lines 17 through 74 are the same as in the previous versions of SimulationController class.

```
28
29          return uniqueInstance;
30      }
31
32      public Properties getProperty() {
33          return props;
34      }
35      public IWeather getWeather() {
36          return weather;
37      }
38      public ISoil getSoil() {
39          return soil;
40      }
41      public IPlant getPlant() {
42          return plant;
43      }
44      public void simulate(Properties props) {
45          // Initializations
46          weather.initialize(props);
47          soil.initialize(props);
48          plant.initialize(props);
49      while (weather.hasNext()) {
50          weather.next();
51          soil.calculateRate();
52          if (plant.isPostPlanting())
53              plant.calculateRate();
54          soil.integrate();
55          if (plant.isPostPlanting())
56              plant.integrate();
57          if (plant.isMature()) {
58              // Stop simulation
```

Figure 9-4. Java implementation of the plug and play SimulatorController (Part 3 of 4).

```
59              plant.setMaturityDay();
60              saveFinalResults();
61              return;
62          }
63        }
64      }
65      private void saveFinalResults() {
66          props = new Properties();
67          props.put("totalPlantDryWeight",
68  new Double(plant.getTotalPlantDryWeight())));
69          props.put("rootDryWeight",
                        new Double(plant.getRootDryWeight())));
70          props.put("fruitDryWeight",
                        new Double(plant.getFruitDryWeight())));
71
            props.put("maturityDay",String.valueOf(plant.getMaturityDay())))
            ;
72      }
73    }
```

Figure 9-4. Java implementation of the plug and play SimulatorController (Part 4 of 4).

Plugging a new class/component into an existing system can be implemented in different ways. One can be organizing the class/component as an executable file that inserts the class/component into the right directory and makes the necessary changes to the configuration file. The next time the system runs, it will activate the newly added class/component. The configuration file can be implemented in several ways. We have selected to use Java's resource bundle, as it is simple and sufficient to demonstrate an example of the plug and play architecture. Other possible implementation for the configuration file is to use an XML file, holding the names of the classes to be created. XML parsers will be needed to extract the information from the XML file and, in combination with the *Builder* pattern [GHJ95], create the required classes.

The advantage of this solution is that it allows development of frameworks for creating complex objects. Information about the interfaces that define the behavior to be implemented by other potential classes can be found in the corresponding UML models. UML allows for creating well-organized and easily-understandable documentation, which can be published on the Web and is available to developers from different groups and locations to coordinate their efforts [PSH04].

5. TESTING UNIT FOR A CLASS/COMPONENT

Testing has always been an important part of the software development. It is the last step of the software development and the step that decides whether the developed system will move to production. The entire system needs to be tested to make sure that it delivers the right results. The use of good modeling practices help to develop good quality software but it may not always avoid mistakes; they are inevitable and part of the process of developing software.

One of the challenges of designing independent components is making sure that they deliver the expected results. Independent components manifest their behavior when involved in a dialog with other components. The more complex the behavior of a component is, the more difficult it is to test. Testing, when done manually, is time consuming. There are several automatic testing methodologies that are successfully used in the software industry. One popular testing software application is JUnit (http://www.junit.org). In this book we will not describe in detail any of the existing testing methodologies. We recommend the readers to look for specialized books in testing methodologies. We will describe some simple testing "patterns" that can be used for testing classes or components.

One way of testing a class or a component in Java is to consider that class as a stand-alone one by adding a static main method and some logic for testing functionalities the class provides. Figure 9-5 shows an example of a main method and code for testing the behavior of class *WeatherDataFromFile*. In this case, we will make sure that the class *WeatherDataFromFile* we have developed provides the right daily weather data needed in the simulation process. The implementation in Java for this class is shown in Figure 8-35 in Section 10.3.1 of Chapter 8.

```
83   static void main(String args[]) {
84       Properties props = new Properties();
85       props.put("weatherFile","weather.txt");
86       WeatherDataFromFile weather = new WeatherDataFromFile();
87       System.out.println(" ---------- Daily Weather Data ------------- ");
88       weather.initialize(props);
89       while (weather.hasNext()) {
```

Figure 9-5. A testing unit for class WeatherDataFromFile (Part 1 of 2).

```
90        weather.next();
91        System.out.println("DayOfYear="+weather.getDayOfYear()
     +" Tmin="+currentDay.getTemperatureMin()
       +" Tmax="+currentDay.getTemperatureMax()
     +" Radiation="+currentDay.getSolarRadiation()
     +" Rainfall="+currentDay.getRainFall());
92        }
93     }
```

Figure 9-5. A testing unit for class WeatherDataFromFile (Part 2 of 2).

Line 83 shows the definition of the method as static and that does not return any results. Lines 84 and 85 create an instance of *Properties* class needed to store the name of the weather data file. Line 86 shows that an instance of the class *WeatherDataFromFile*, referred to as weather, is created. Line 87 prints the string "Daily Weather Data". Line 88 initializes the object weather using the parameter "weather.txt," which is the file name holding the weather data to be used in the simulation process. By definition, class *WeatherDataFromFile* implements interface *Iterator*. Therefore, it provides the means to iterate over the weather data. Line 89 tests whether the end of the file is reached. Line 90 obtains an object of type *DailyWeatherData* that is referenced by attribute *currentDay*. From the object *currentDay*, daily data can be obtained by sending messages such as *getTemperatureMin()*, *getTemperatureMax()*, etc. Line 91 prints daily weather data for the current day.

Figure 9-6 shows a partial view of the weather data read from the file weather.txt. Notice that these results are the same as the ones shown in Figure 8-36. Therefore, class *WeatherDataFromFile* provides the expected results.

```
-------------- Daily Weather Data ----------------
DayOfYear=1  Tmin=5.1  Tmax=20.0  Radiation=4.4  Rainfall=23.9
DayOfYear=2  Tmin=10.8 Tmax=13.3  Radiation=1.1  Rainfall=0.0
DayOfYear=3  Tmin=12.1 Tmax=14.4  Radiation=1.1  Rainfall=0.0
DayOfYear=4  Tmin=3.6  Tmax=18.3  Radiation=6.1  Rainfall=14.7
DayOfYear=5  Tmin=12.8 Tmax=17.2  Radiation=5.6  Rainfall=0.8
DayOfYear=6  Tmin=12.4 Tmax=21.1  Radiation=5.0  Rainfall=0.0
DayOfYear=7  Tmin=11.1 Tmax=21.7  Radiation=7.2  Rainfall=0.0
DayOfYear=8  Tmin=12.0 Tmax=21.7  Radiation=8.3  Rainfall=0.0
DayOfYear=9  Tmin=6.1  Tmax=20.0  Radiation=8.3  Rainfall=0.0
DayOfYear=10 Tmin=3.5  Tmax=23.3  Radiation=11.1 Rainfall=3.8
DayOfYear=11 Tmin=13.5 Tmax=19.4  Radiation=2.8  Rainfall=0.0
```

Figure 9-6. Daily weather data obtained from the testing unit.

In a similar manner, we will create a testing unit for the class *SimulationController*. The class definition for *SimulationController*, as defined in Figure 8-39, will be provided with a static method referred to as *main* as shown in Figure 9-7.

Lines 72 through 75 define the parameters needed for the simulation process and assign them initial values. Usually, these values will be provided by users of the system. Line 76 obtains the unique instance of the *SimulationController* and line 77 sends *simulator* the message *simulate* with the appropriate parameters. The results of the simulation are stored in a *Properties* file by the simulator.

```
70  static void main(String args[]) {
71      Properties props = new Properties();
72      props.put("weatherFile","weather.txt");
73      props.put("plantingDate","121");
74      props.put("soilDepth","145");
75      props.put("wiltingPointPercent","0.06");
76      SimulationController
                  simulator = SimulationController.getInstance();
77      simulator.simulate(props);
78      System.out.println("----- Simulation Results -----");
79      Properties pr = simulator.getProperty();
80      Double rdweight = (Double)pr.get("rootDryWeight");
81      Double ptdweight = (Double)pr.get("totalPlantDryWeight");
82      Double fdweight = (Double)pr.get("fruitDryWeight");
```

Figure 9-7. Test unit for class SimulationController (Part 1 of 2).

```
83        String matDay = pr.getProperty("maturityDay");
84        System.out.println("Root Dry Weight="+ rdweight.toString()
85                +" Maturity day="+ matDay
86                +" Total Plant Dry Weight="+ptdweight.toString()
87                +" Fruit Dry Weight="+fdweight.toString());
88     }
```

Figure 9-7. Test unit for class SimulationController (Part 2 of 2).

Lines 80 through 83 read the results from the *Properties* file and lines 84 through 88 print the results. The obtained results can be evaluated by a specialist to make sure they are accurate, or they can be compared to results obtained with other versions of the software if they exist.

The ability to test independently class/components allows for a faster software development as the focus is on one class/component with a well-defined behavior. By integrating testing unit into the build process, it makes it easier to discover implementation errors and design flaws. A class that has successfully passed the individual test can be easily integrated into more complex testing scenarios where several classes/components are involved. Thus, using an iterative process that consists of testing individual classes and later the interaction of several classes, makes it easier to test complex software. We have used this approach all along the software development process and have obtained good results.

Chapter 10

SOIL WATER-BALANCE AND IRRIGATION-SCHEDULING MODELS: A CASE STUDY

1. INTRODUCTION

In this chapter we will discuss issues related to the development of a general UML model that covers a large class of similar models, the class of water-balance and irrigation-scheduling models [PSH04]. Many irrigation-scheduling and water-balance models have been developed and published in the past. These models have been used for both research purposes and as management tools. Models used for research purposes generally represent the system and underlying processes in greater detail than do management models. [AW85] distinguished between (i) mechanistic and functional, and (ii) rate and capacity models. Mechanistic models are based on fundamental processes, whereas functional models simplify the representation of processes. Rate models are driven by time and define rates of change within a system; capacity models are driven by input amounts and define amounts of change. However, even within these broad categories, models differ in their assumptions and representation of water-balance processes. [MLB98] make a detailed analysis of the assumptions and the representations of the water-balance models. Water-balance models have been used as stand-alone applications and as components of larger agricultural-system models. For example, a water-balance model developed by Ritchie has been integrated into numerous simulation models, including the cotton simulation model OZCOT [Hea94], CERES-Wheat [RO85], and is used by the Decision-Support System for Agrotechnology Transfer (DSSAT) [Rit98] which includes crop simulation models for a number of agronomic crops. Irrigation-scheduling models are generally standalone applications that have been designed as management decision-support tools.

Similar models may differ in their input data requirements and their use. [OEK01] used THESEUS developed by [Weg00], a modeling system containing a number of sub-models, for water-balance and crop simulation, representing the soil, plant, and atmosphere, which can be combined to create simulation models. The system contains a number of water-balance models; users can select one that meets their complexity and data requirements. They summarized and distinguished between models according to their output, the equations used in the model and the input data requirements.

Although water-balance and irrigation-scheduling models may have been developed for different purposes and vary widely in their input requirements and representation of processes, they do share a number of commonalities. For example, they all typically require some soil and weather data. There is also often overlap in the processes represented, although they may be calculated by different methods. For example, most models include water removal by evapotranspiration. This may be calculated by the model: A historical value or an input requirement. The process of water movement is simulated by these types of models either as amounts moving into the soil profile and stored within it, or by rates of change in soil water content. In order to identify common elements and relationships, a number of water-balance and irrigation-scheduling models were compared.

In the case of soil water-balance and irrigation-scheduling models, the common system elements are the soil, plant, and weather. The behavior of these elements is model-specific and is defined by the processes accounted for by the model.

2.   CONCEPTUAL MODELS: EXAMPLES

As previously mentioned, the most common elements used in water-balance and irrigation-scheduling models are plant, soil, and weather. A first draft of the conceptual diagram is presented in Figure 10-1.

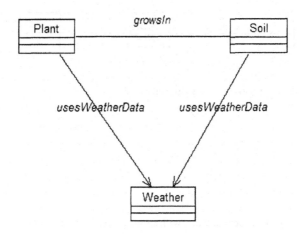

Figure 10-1. Conceptual model for water-balance and irrigation-scheduling models.

As shown in Figure 10-1, *Plant* and *Soil* are linked with a bidirectional association referred to as *growsIn*. This association shows that a plant grows in soil and takes from soil all the nutrients that are needed for plant development; the plant consumes nutrients that are located in soil. Processes occurring in *Plant* need *Soil* data and vice versa; processes occurring in *Soil* need *Plant* data. Both, *Plant* and *Soil*, are connected via a unidirectional association to *Weather*, referred to as *usesWeatherData* that expresses the fact that plant and soil are affected by the weather conditions.

The conceptual model presented above does not take into consideration the fact that additional water is provided by the means of irrigation when there is a drought for a considerable amount of time. Therefore, another element needs to be added to the conceptual diagram: The *irrigation management* element. Figure 10-2 shows the modified conceptual diagram.

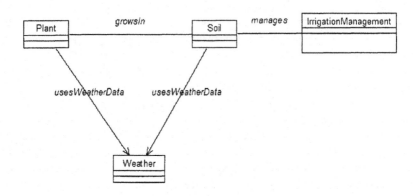

Figure 10-2. Conceptual model with irrigation management considerations.

Figure 10-2 shows that irrigation management practices will be applied to soil and plant in order to improve production yield. The association *manages* that links *Soil* and *IrrigationManagement* is bidirectional, meaning that processes occurring in soil need irrigation management data and calculations occurring in the management need soil data. It is important to note that it is not always easy to define the nature of the associations among the elements of the model. As we are yet in the phase of developing a conceptual model, not all the exact relationships between elements can be defined at this point. The more we advance in the model construction, the more we will know about the collaboration between elements of the model and the more precise the model becomes. As an example, some association that we have considered as unidirectional may become bidirectional as we learn that data from the class located in one side of the association are needed to calculate processes in the class in the other side of the association. The process of developing a complete, detailed, and exact model is an iterative one.

The calculation of some processes requires data from several elements of the conceptual model and cannot be assigned to a particular element. Such processes are the calculation of evapotranspiration rates; they usually need data from soil, plant, and the weather elements. Therefore, an additional element needs to be added to the conceptual model to carry out these calculations. This element, referred to as *SoilPlantAtmosphere*, will be assigned the task of performing these calculations and is added to the conceptual diagram as shown in Figure 10-3.

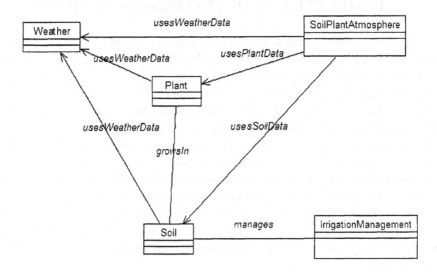

Figure 10-3. Conceptual model that considers soil-plant-atmosphere processes.

As shown in Figure 10-3, a new element, the *SoilPlantAtmosphere*, is added to the conceptual diagram to account for evapotranspiration rate calculations. In order to perform these calculations, *SoilPlantAtmosphere* needs to have access to other elements containing these data. Thus, the unidirectional association *usesWeatherData* between *SoilPlantAtmosphere* and *Weather* allows access to weather data. The association *usesPlantData* allows for accessing plant data and the association *usesSoilData* allows access to soil data.

The conceptual diagram presented in Figure 10-3 seems to cover all the elements and their relationships needed to calculate processes that occur in soil, plant, and atmosphere that are required to calculate the amount of water needed for irrigation. The relationships between elements are independent of the type of equations used in the model and of the programming language used for the implementation of the model. A general structure, representing all the elements involved in a water-balance or irrigation-scheduling model and their relationships is developed to serve as a general template for developing new models.

3.   TEMPLATE FOR DEVELOPING NEW MODELS

The plant, weather, and soil plant atmosphere elements of the conceptual model are presented by the classes *Plant*, *Weather*, and *SoilPlantAtmosphere*, as shown in Figure 10-4. Soil is represented by three classes; *SoilProfile*, *SoilLayer*, and *Cell*. The *SoilProfile* class represents the profile as a whole. Some models consider soil as a composition of layers with different properties. Some other models consider the entire soil profile as one layer. The *SoilLayer* class is created to store layer-specific data and behavior. The *Cell* class represents the surface. It represents a uniform area for which a simulation is run. The *GroundWater* class is created to store data and behavior for the groundwater layer, sometimes considered by water-balance and irrigation-scheduling models. The *IrrigationManagement* class plays an important role in irrigation-scheduling models. It stores information related to irrigation management practices and calculates outputs such as recommended irrigation rates.

A unidirectional association, referred at as *usesWeatherData*, links classes *SoilPlantAtmosphere* and *Weather*. The navigation direction is from *SoilPlantAtmosphere* towards *Weather*. This means that objects of class *SoilPlantAtmosphere* have access to data and behavior to objects of class *Weather*. Weather data are required to calculate processes occurring in *SoilPlantAtmosphere*, such as calculations of actual evapotranspiration rates. Objects of class *Weather* do not have access to objects of class *SoilPlantAtmosphere*, as there are no calculations occurring in this class. The multiplicity of the association *usesWeatherData* is one-to-one, meaning that one object of type *SoilPlantAtmosphere* has access to one object of type *Weather* (one source of weather data) at the time of the simulation.

The association *usesWeatherData* that links classes *Cell* and *Weather* has the same properties as the one linking classes *SoilPlantAtmosphere* and *Weather*; it is bidirectional and a one-to-one association.

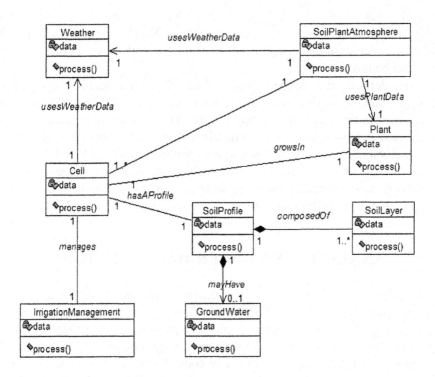

Figure 10-4. Template for developing new water-balance or irrigation-scheduling models.

The association between classes *SoilPlantAtmosphere* and *Plant* is unidirectional, with the navigation direction from *SoilPlantAtmosphere* towards *Plant*. An object of type *SoilPlantAtmosphere* can access data and behavior from an object of type *Plant*. The vice versa is not true; an object of type *Plant* cannot access data and behavior from an object of type *SoilPlantAtmosphere*. Processes occurring in class *SoilPlantAtmosphere* need plant data for their calculations. The multiplicity of the association is one-to-one; one object of type *SoilPlantAtmosphere* can access one object of type *Plant*.

The association between classes *SoilPlantAtmosphere* and *Cell* is bidirectional, as processes occurring in each of the classes need data from the other class. The multiplicity of the association is one-to-many; one object of type *SoilPlantAtmosphere* can access one or more objects of type *Cell*.

In the template shown, *Cell* is the central unit of the system. It has a soil profile associated with it, which, in turn, is composed of one or more layers. A soil profile may also contain groundwater. *Plant* grows in a *Cell* and the two classes are able to exchange data. Any soil data required is accessed through the *Cell*. Class *Cell* does have access to data from *SoilPlantAtmosphere*, which it is able to pass to *SoilProfile* and *SoilLayer*, in order that, for example, water removal from the soil by evapotranspiration (ET) can be modeled.

After defining the elements involved in irrigation scheduling and water-balance models, attributes and behavior are defined for each element represented in the diagram. Attributes represent the input data required by the model and behavior defines the processes that are calculated. Figure 10-4 shows elements, represented by classes, provided with data and behavior. Based on the general template, class diagrams were developed for two models: A water-balance model and an irrigation scheduling model that will be presented in the following sections.

4.   ANALYSIS OF A WATER-BALANCE MODEL

In order to develop specific class diagrams for individual models, each participating class in the template is populated with attributes and methods representing the model-specific input requirements and processes. This structure clearly shows the model data input requirements and the processes represented and can be used to organize code. Figure 10-5 represents a class diagrams for a water-balance model developed by Ritchie [Rit98]. Ritchie's model requires certain weather data, such as rainfall, solar radiation, and minimum and maximum temperature; therefore, class *Weather* has been provided with the attributes to store those corresponding values. Class *Weather* is provided with a method named *calculatePriestleyTaylorET*, which is used to calculate reference ET. Depending on the method used, calculation of actual evapotranspiration may involve weather, soil, and plant data. Therefore, class *SoilPlantAtmosphere* has links to *Weather*, *Soil*, and *Plant* classes. Reference ET is calculated solely from weather data; this provides a basis for the estimate of evapotranspiration. This reference ET, is normally then modified to determine the actual evapotranspiration. Actual ET is determined by the crop characteristics and by the soil moisture content. Different crops have different water requirements, which vary temporally, and the actual amount of water that can be physically removed from the soil is limited by the amount of water actually stored in the soil. Methods of calculating actual ET from reference ET may use crop information, soil data, or both. Therefore, it is necessary for class *SoilPlantAtmosphere* to have relationships with classes *Cell*, *Weather*, and *Plant* to access to all these data. Unlike estimates of actual ET, reference ET calculations, such as the method *calculatePriestleyTaylorET* in Figure 10-5, were assigned to class *Weather*.

Reference ET is determined solely from weather data and represents potential evapotranspiration rates based on weather conditions, rather than any processes occurring in the plant or soil or interactions between the two. Online weather systems, like the Florida Automated Weather Network (FAWN) [BJ98] or MetBroker [LKN02], sometimes provide already-calculated reference ET values. If these systems are used, reference ET could be imported from the data provider system.

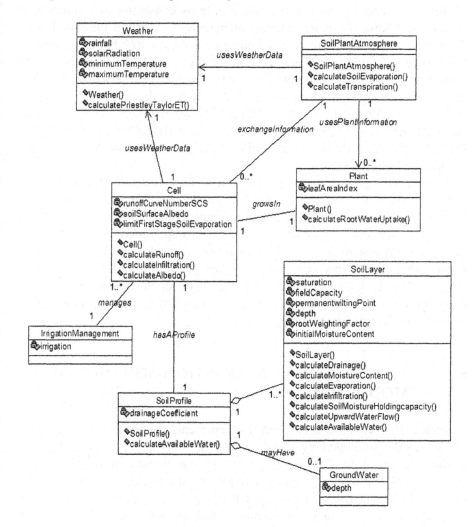

Figure 10-5. Class diagram for Ritchie's model.

The class *SoilPlantAtmosphere* is provided with behavior to determine the water loss through evaporation and transpiration. Ritchie's model separates these processes, considering evaporation and transpiration separately. Therefore, class *SoilPlantAtmosphere* is provided with two methods: *calculateSoilEvaporation* and *calculateTranspiration*. Soil evaporation and transpiration are calculated using soil moisture data and leaf area index is required in the calculation of transpiration.

Ritchie's model considers the soil profile as consisting of a number of layers and allows the representation of numerous layers in the simulation. Thus, the multiplicity of the association *composedOf* between classes *SoilProfile* and *SoilLayer* is one-to-one-or-many. This model requires a great number of layer data and it calculates many layer-related processes because the water movement into and out of layers is considered and modeling of water use is detailed, being partitioned between layers. Processes and attributes pertaining to the profile as a whole are assigned to class *SoilProfile*. Ritchie's model uses a whole-profile drainage coefficient stored in the attribute *drainageCoefficient*.

Class *Cell* is provided with data and behavior to calculate the amount of water that enters the soil surface. Ritchie's model uses The United States Soil Conservation Service (SCS) to determine run off and, in turn, calculate the infiltration amount. Therefore, class *Cell* is provided with attributes *runoffCurveNumberSCS*, *soilSurfaceAlbedo*, and *limitFirtsStageSoil Evaporation* and methods *calculateRunoff*, *calculateInfiltration*, and *calculateAlbedo*. Ritchie's model uses the depth of the groundwater layer; therefore, class *GroundWater* is created and provided with the attribute *depth* to store the corresponding value.

5.   ANALYSIS OF AN IRRIGATION-SCHEDULING MODEL (ISM)

Figure 10-6 contains a diagram that represents the Irrigation Scheduling Model (ISM) developed by [GSR00]. As in the Ritchie model, the only methods assigned to class *Weather* are for calculation of reference ET. However, ISM provides several methods for its calculation, such as the Penman–Monteith, Blaney Criddle, or Priestley–Taylor methods. The model also allows for user input of already-calculated reference ET. The ISM *Weather* class has been provided with a larger number of attributes, compared to that of the Ritchie model. These extra attributes are needed to store the data values required by each of the methods of calculating reference ET. Not all the input data need to be supplied to the model - only those required by the selected calculation method. Class *Plant* in ISM has a richer set of attributes than Ritchie's model. This illustrates a difference between the two models. Whereas Ritchie's model requires the input of leaf area index (either from the user or from a linked model), ISM includes processes that simulate aspects of crop growth. As a result, much more information about the plant is required by ISM. Choice of a model may be, in part, determined by data availability. For example, use of ISM may be preferable if the leaf area index required by Ritchie's model cannot be readily obtained. In this case, corresponding data inputs required by ISM may have to be obtained to allow the model to make predictions about the condition of the crop.

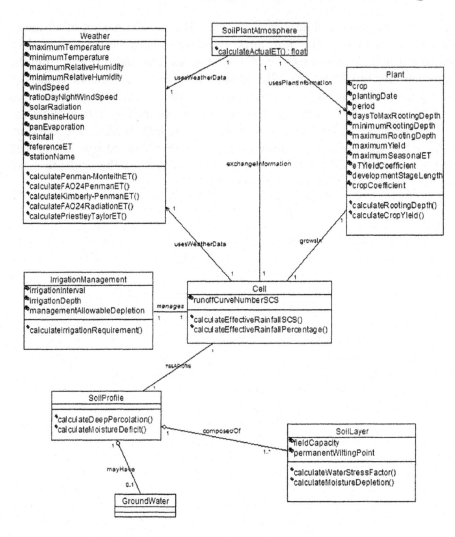

Figure 10-6. Class diagram for the ISM model.

Class *SoilPlantAtmosphere* is provided with data and behavior for determining water losses through evaporation and transpiration. The ISM model predicts evapotranspiration by combining the effects of both mechanisms to determine overall water loss from the plant-soil system. ISM converts reference ET values, provided by class *Weather*, into actual evapotranspiration based on soil moisture availability, provided by class *SoilLayer*, and the crop coefficient, provided by class *Plant*. The need for data from *Weather*, *Plant*, and *SoilLayer* justifies the associations class *SoilPlantAtmosphere* has with *Weather*, *Plant*, and *Cell*. Although there is no direct association between *SoilPlantAtmosphere* and *SoilLayer*, the communication between these two classes takes place via *Cell*.

Classes *Cell, SoilProfile, SoilLayer,* and *GroundWater* are modeled to function together as a component and could be considered as a *Soil* component. The main class in this component is class *Cell*, which plays the role of the gatekeeper. One may suggest that a direct link from class *SoilPlantAtmosphere* to class *SoilLayer*, bypassing class *Cell*, would shorten the communication between these two classes. The problem with this solution is that by shortening the path, we will increase the interdependency between classes of the system. In the case that some other *Soil* component providing the same behavior can be found, it will be difficult to use it, as our system will not allow the substitution of one group of classes with another one. Having class *Cell* controlling the dialog of the soil related classes with the rest of the system makes it easy to define an interface that provides the services the component can offer. An interface allows substitutability between different components implementing the same interface.

Layer-specific data and behavior are assigned to class *SoilLayer*. Unlike the Ritchie's model, ISM does not partition the soil into layers; it simply considers the soil profile as a single layer that extends to the bottom of the root zone. The association *composedOf* between classes *SoilProfile* and *SoilLayer* is one-to-one-or-many; therefore, the single layer approach of the ISM model is taken into consideration. Class *SoilLayer* is provided with a smaller number of attributes than the same class in the Ritchie's model, because the single layer approach requires fewer parameters for calculation of the water movement. Processes and attributes that apply to the soil profile as a whole are assigned to class *SoilProfile*. ISM model calculates deep percolation out of the root zone; thus, class *SoilProfile* is provided with a method referred to as *calculateDeepPercolation*.

Class *Cell* is provided with data and behavior for calculating the amount of water that enters the soil surface. The ISM model describes this as effective rainfall and uses the SCS method to determine this amount. The method *calculateEffectiveRainfallSCS* implements the SCS method for this calculation. Alternatively, a fixed percent of actual rainfall can be specified. ISM includes a number of irrigation-scheduling functions in class *IrrigationManagement*. The user can input either a desired irrigation interval or an allowable depletion value and the model calculates an irrigation requirement to guide the management. The ISM model does not take into consideration the groundwater; therefore, the *GroundWater* class does not provide any data or behavior.

6.   THE BENEFITS OF A GENERAL TEMPLATE

Templates can facilitate the development of software construction. The general template shown in Figure 10-4 represents a template for building a new water-balance or irrigation-scheduling model. It represents the common classes used for these types of models and their relationships. Each of the classes of the template can be populated with attributes and methods needed by the particular approach used for model development, as shown in Figures 10-5 and 10-6.

In addition to the documentation provided by the class diagram, UML tools usually include methods for creating complementary documentation. The software used allowed users to create extra documentation for each component of the class diagram: The relationships, classes, class attributes, or class methods. This extra documentation allows the developer to explain the meaning and the role of each of these in the system. Attributes can be defined and equations used in the calculation of processes can be presented, for example, in pseudo-code. That is, equations can be presented in a format similar to that used in papers, books, etc. It need not be written in a specific programming language. An example of documentation for a method is shown in Figure 10-7.

The general template could easily be extended to accommodate other agricultural and environmental models. For example, models of crop nutrient uptake rely on the same basic system elements as soil water-balance models. They could be represented with little or no modification of the template, perhaps requiring a class to represent additional management practices such as fertilizer management. Ultimately, the template could be extended to include numerous classes representing all aspects of such systems.

From these templates, UMLs forward-engineering capabilities can be used to generate skeleton programming code, which can form the basis of the final code implementation. Usually, UML tools provide a means of translating diagrams into several implementation languages such as Java, Visual Basic, C++, etc. Rational Rose (http://rational.com) provides capabilities for code generation in a variety of languages such Java, Visual C++, Visual Basic, ADA83, ADA95, CORBA, and XML_DTD. This capability greatly simplifies the final software development process. Figure 10-8 shows an example of automatic code generation in Java for class *Cell*. Figure 10-9 shows an example of automatic code generation in Visual Basic for class *SoilLayer*, and Figure 10-10 shows an example of code generation for class *SoilPlantAtmosphere*. In Figure 10-10, the code generation includes the relationships class *SoilPlantAtmosphere* has with classes *Plant* and *Weather*.

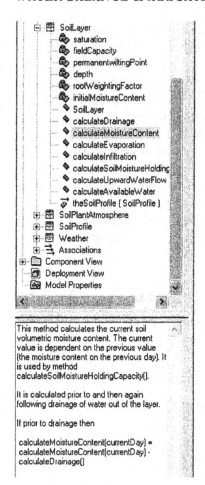

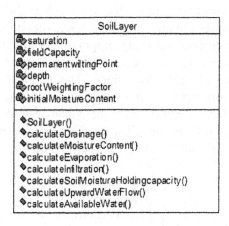

Figure 10-7. Example of documentation provided for a method.

Class diagrams created with UML and Rational Rose (other software provide publishing capabilities as well) can be used to create well-organized and easily understandable documentation, which can be published on the Web. Publishing the diagrams on the Web allows developers from different groups, in different locations, to easily coordinate their efforts. Additionally, because UML diagrams use plain English and can be understood by programmers and non-programmers alike, they can facilitate collaboration between these groups. Model-specific documentation for existing and future models can be created using the methodology presented and can aid in model evaluation. The template developed can be used to represent various models regardless of the approach used in the model, using a common set of visual elements. Both single and multi-layer approaches to modeling changes in soil

water content could be described using the same UML diagram and the
template has the potential to be expanded to represent other soil–crop system
models, such as nutrition models. In addition to documenting the underlying
science of the models represented, the template outlines the structure of the
code organization and can be used to build new models. Generating skeleton
code from class diagrams, using UML tools, can greatly simplify code
implementation. Models could be implemented in different programming
languages, allowing programmers to choose the best environment according
to their specific architectural needs or to update models in newer
programming languages. The template could be used to facilitate model
maintenance and modification, allowing the programmer to easily locate
where changes must be made. Subsequent reuse of code could also be
simplified. The following figures show examples of code generation in Visual
Basic and Java using the same class diagram.

```
1   public class Cell {
2       private int runoffCurveNumberSCS;
3       private double soilSurfaceAlbedo;
4       private double limitFirstStageSoilEvaporation;
5       private SoilProfile theSoilprofile;
6       public Weather the Weather;
7       public Plant thePlant;
8       public SoilPlantAtmosphere theSoilPlantAtmosphere;
9       public Cell() {}
10      public void calculateRunoff(){}
11      public void calculateInfiltration(){}
12      public void calculateAlbedo(){}
13  }
```

Figure 10-8. Example of code generation in Java.

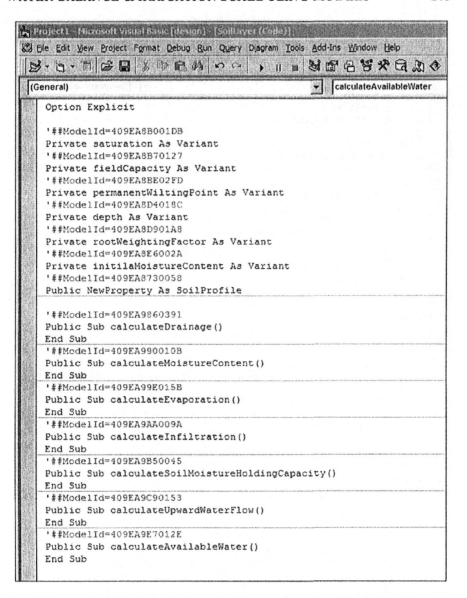

Figure 10-9. Example of code generation in Visual Basic for class SoilLayer.

In Visual Basic, operations are defined as subroutines starting with the keywords **Public Sub** and ending with **End Sub**. Between these words, the user will enter the logic for the operation.

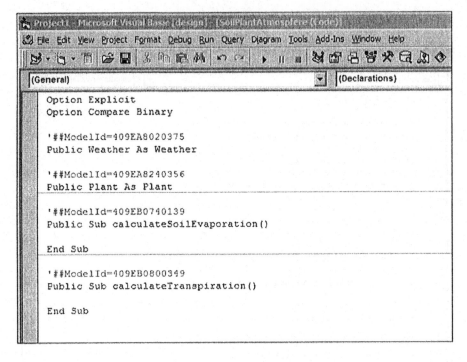

Figure 10-10. Example of code generation in Visual Basic for class SoilPlantAtmosphere.

Chapter 11

DISTRIBUTED MODELS

1.   INTRODUCTION

Usually, the environmental models, regardless of the chosen implementation environment, are designed to run on a single computer. This means that all objects (in case of object-oriented languages) or variables (in case of traditional programming languages) reside in the same machine and share the same address space. This way of developing software is called local computing [WWW94]. In recent years, the number of applications that use parts or elements residing on different computers has increased considerably. This way of developing software is referred to as distributed computing [WWW94]. There are several advantages to distributed computing. One is simplifying maintenance. A component shared by different distributed applications can reside in one location, thus giving ownership and control to the institution that creates and maintains the component. It is not necessary to distribute copies of this component that would be difficult to track and update. Another advantage is economic. Because of the high cost of hiring experienced programmers, more and more companies prefer outsourcing the software development rather than developing it at home. The need for "off-the-shelves" solutions is obliging companies to give serious considerations to the interoperability between existing and newly developed software [Bro99]. In distributed computing, little information is known about the object receiving the message. There is no information about the operating system or the hardware architecture in which the receiving object is running, nor the implementation language used. It is only known that the receiving object implements a certain interface and the message is part of the definition of this interface. There are several technologies that can be used to develop distributed applications. The dominant technologies are: The Object Management Architecture (OMA) of Object Management Group (OMG) (http://www.omg.org), Microsoft's distributed computing architecture (http://microsoft.com/), and Sun's Java-based distributed component technology (http://java.sun.com/). The following is a brief description of these approaches for developing distributed component-based applications.

2.   CORBA

OMG was created in 1989 to develop, adopt, and promote standards for the development and deployment of applications in distributed heterogeneous environments [Vin97], [VD98]. OMG is a large consortium with more than 800 companies trying to reach a consensus on an appropriate component model and services for developing component-based distributed applications (http://www.omg.org). OMG is the world's largest computer consortium and is a nonprofit organization that started initially with eight members: 3Com, American Airlines, Canon, Data General, Hewlett-Packard, Philips Telecommunications N.V., Sun Microsystems, and Unisys [VD98]. OMG does not develop any technology per se, nor does it advertise any product of its members. OMG's goal is to promote the object-oriented approach to software engineering and create a general architectural framework for developing component-based distributed applications based on the interface specifications for objects of the application. OMG works to provide standards for building component-based applications and encourages its members to follow these standards.

The component-based approach is the heart of OMG's Object Management Architecture (OMA). OMA defines the specifications for the underlying distributed architecture and the way components dialog with each other in a distributed environment. OMA is the general framework that embraces all technologies adopted by OMG [VD98].

The Common Object Request Broker Architecture (CORBA) is one the most important middleware project undertaken by the software industry. It is OMG's response to the challenging problem of building communication bridges between isolated islands developed in different programming languages and computing platforms. CORBA enables natural interoperability, regardless of platform, operating system, programming language, and even of network hardware and software. CORBA defines a mandatory TCP/IP-based protocol for interoperability over the Internet and most intranets. CORBA clients can run on a large variety of computers, from hand-held wireless palmtops or pagers to desktop machines or mainframes. CORBA servers can also run on all these machine types. The specification standardizes complex resource management and fault tolerance for large, reliable server-side applications. There are also specialized versions of CORBA for real-time and small embedded servers. There is strong support for CORBA on the application side by the OMG, a collection of standardized objects performing functions, including the key enterprise-required services for transaction handling and security. CORBA goes beyond this to define standard objects and frameworks in business domains, such as finance, insurance, manufacturing, the health sector, and more. CORBA is supported by UML that is OMG's standard for Object-Oriented Analysis and Design.

CORBA is composed of three parts: A set of interfaces that can be invoked by users, the object request broker (ORB), and a set of object adaptors [Szy99].

2.1 The Interface Definition Language (IDL)

Interfaces need to be described in a common language [VD98]. This common language is referred to as Interface Definition Language (IDL) and is used to describe the interface of an object. IDL is an object-oriented declarative language for specifying server interfaces. It is not a programming language; it cannot be used to write code. As we know, interfaces define the operations that an implementation object should provide. As an example, Figure 11-1 shows the module definition for a hypothetical environmental model. Line 1 defines the module *EnvironmentalModel*. Modules provide a name scope for identifiers in an IDL specification. This scope prevents name clashes for identifiers defined in other modules. As for example, within the module *EnvironmentalModel*, only one interface, referred to as *IWeather* can, be declared. Line 3 defines interface *IWeather* within the scope of module *EnvironmentalModel*. Lines 4 through 12 define methods for interface *IWeather*. Line 14 defines interface *ISoil* within the scope of module *EnvironmentalModel*, and line 15 defines the only method of interface *ISoil*. The interface definition of object *IWeather* shows that the object should provide weather data, such as solar radiation, average temperature, minimum temperature, maximum temperature, rainfall, etc. Any class that implements this interface would specify how these weather data would be obtained.

```
1   module EnvironmentalModel {
2
3   interface IWeather {
4       public double getSolarRadiation();
5       public double getAverageTempDuringDay();
6       public double getAverageTemperature();
7       public double getAverageTempForET();
8       public double getRainfall();
9       public double getTemperatureMin();
10      public double gettemperatureMax();
11      public double getPotentialET();
12  };
13
14  interface ISoil {
15      public double getWaterStress();
16  };
17 };
```

Figure 11-1. Example of an IDL interface definition for components Weather and Soil.

Java maps IDL modules to a package with the same name and every package corresponds to a directory in the file system. The IDL compiler creates a directory for each module and all generated files are saved in this directory.

As CORBA has to work with different programming languages, there are mapping methods that map types and structures from the common language, IDL, to a number of programming languages. For example: The boolean type in Java is mapped to boolean type of IDL, the byte type in Java is mapped to octet type of IDL, etc [Vin97]. Currently, there are mappings to IDL for Java, C, C++, COBOL, Smalltalk, ADA, etc. (http://www.omg.org).

2.2   The Object Request Broker (ORB)

The Object Request Broker (ORB) is the heart of the system. The ORB acts as a central Object Bus over which each CORBA object interacts transparently with other CORBA objects located either locally or in other remote servers. The ORB is responsible for finding a CORBA object's implementation, preparing it to receive requests, communicating requests to it, and carrying the answer back to the client. The ORB establishes a communication bridge between the client and the server, as shown in Figure 11-2. The underlying ORB implementation is not relevant to distributed system developers. Once the communication bridge is established, the client can access data and behavior from the object residing in the server as it were a local object.

Furthermore, the client's interface is completely independent from the implementation residing in the server. All that the client needs to know is the set of interfaces the object residing in the server offers. The interfaces to the ORB and to objects built using the ORB are well defined, providing a uniform framework across the entire distributed environment and making applications built using an ORB very portable across diverse platforms.

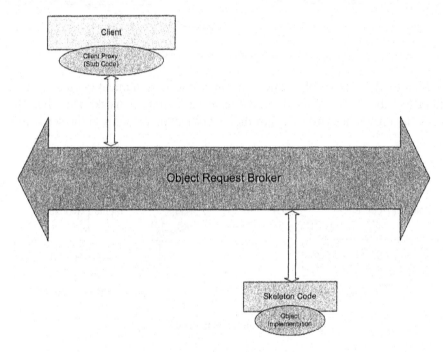

Figure 11-2. Object request Broker links clients and server applications.

An ORB delivers requests from client applications to server applications. This process occurs in three steps [Ros98], as shown in Figures 11-3, 11-4, and 11-5.

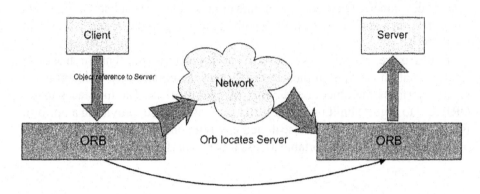

Figure 11-3. The client has a reference to the object server.

Figure 11-3 shows that ORB manages the communication between the client and the server; ORB locates the server in the network. The client has access to a reference (or proxy) of the real object that resides in the server.

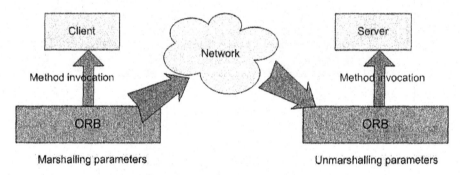

Figure 11-4. Client refers to the proxy object.

Once the communication is established between the client and the server, the client sends messages to the proxy as it were the real object, as shown in Figure 11-4. The message may include parameters needed for the execution of the method. ORB marshals these parameters, meaning it converts these parameters into a format that can be transmitted across the network to the remote object. The receiving server unmarshals the message and executes the method.

ORB is also in charge to bring to the client the results of the method's execution, as shown in Figure 11-5. To do so, ORB marshals the results and ships them to the client that unmarshals them into an understandable format for the client. Note that the process of marshalling/unmarshalling the message and the parameters is transparent to the user. The entire process is silently managed by the object request broker (ORB).

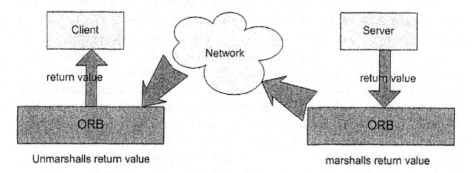

Unmarshalls return value   marshalls return value

Figure 11-5. ORB sends back to client the answer from server.

2.3   Adaptors

The CORBA standard describes a number of object adapters; with the task of interfacing an object's implementation with its ORB. The OMG provides three sample object adapters: The Basic Object Adapter (BOA), which is the most used one, the Library Object Adapter, and the Object-Oriented Database Adapter. The last two adaptors are mainly used for accessing objects in persistent storage.

The BOA provides CORBA objects with a set of methods for accessing functions defined in ORB. These functions range from user authentication to object activation to object persistence. The BOA is the CORBA object's interface to the ORB. BOA plays a slightly different role in the server application compared to the role it plays in the client application. In the server implementation, BOA informs the ORB when objects are ready to receive incoming requests. In the client application, BOA is the component of ORB that makes sure that the reference or the proxy reaches the real object located in the server. The next two sections will provide examples of the use of adaptors in server and client applications. According to the CORBA specification, the BOA should be available in every ORB implementation, and this seems to be the case with most CORBA products available in the market.

In the next sections we will present a simple example of implementation of a CORBA soil server and a soil client.

2.4   A CORBA Soil Server

Let us consider the Soil component developed in Chapter 8, **The Kraalingen Approach**, and implement a simplified version of it as a server using CORBA. This means that the soil class/component will provide its services to any other class/component or system residing in the same address space or in other servers located anywhere in the network. Therefore, the behavior of this class/component will be slightly different; besides the behavior needed in the simulation process, the component will be provided with additional behavior to function as a CORBA server. For simplicity reasons, we will provide soil component only with the behavior needed to provide information about its soil depth and wilting point values.

One of the first tasks the server should accomplish is to obtain references to ORB and BOA. After which, ORB and BOA objects will be created and the server makes them available for use to any other object outside the system. Then, the server executes a dispatch loop to wait for other objects invoking its services. Clients must connect to the object residing in the server to use its services.

Although CORBA interfaces are defined in a very general manner, different vendors, while respecting the general definition of interfaces, have implemented their behavior slightly differently. Therefore, the implementation of a CORBA service may be different in different products. The most well-known commercial vendors of CORBA technology are BORLAND's Visibroker (http://www.borland.com/visibroker/) and IONA's ORBIX (http://iona.com).

The Integrated Development Environment (IDE) used to implement the examples is Borland's JBuilder (**http://borland.com**) which uses Visibroker technology for CORBA. The use of the Visibroker technology implies some particularities that make this implementation slightly different from the cases where other CORBA technologies (such as IONA's ORBIX, http://iona.com) are used. Figure 11-6 shows the class diagram for the *SoilCORBA* server.

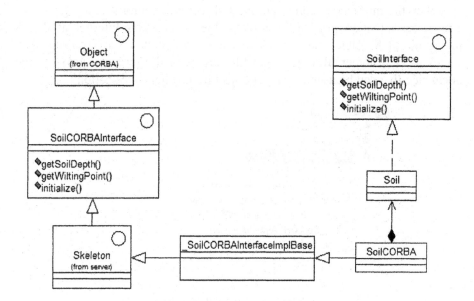

Figure 11-6. Class diagram for SoilCORBAServer.

To implement the *SoilCORBA* server, first its interface *SoilCORBAInterface* should be defined using CORBA's Interface Definition Language (IDL), as shown in Figure 11-7. The interface will define the functionalities that this server will provide remotely to any client. The interface *SoilCORBAInterface* is similar to *SoilInterface*, as shown in Figure 11-6. The only difference is that *SoilCORBAInterface* provides capabilities for sending/receiving messages through the Internet. Therefore, *SoilCORBAInterface* inherits the required behavior from CORBA's main Object that is part of CORBA's arsenal for moving objects across the Internet, as shown in both Figures 11-6 and 11-7. Every interface inherits from CORBA.Object.

```
1    public interface SoilCORBAInterface extends
            org.omg.CORBA.Object {
2       public double getSoilDepth();
3       public double getWiltingPoint() ;
4       public void initialize();
5    }
```

Figure 11-7. Interface definition for class SoilCORBA.

When the interface is compiled, the IDL compiler automatically generates a few classes that are needed for moving objects across the Internet, as shown in Figure 11-8. The generated classes are divided in three groups: stubs, skeletons, and helper files. The stub files are needed to create a proxy in the client side that can be used as a local object in the client application.

Figure 11-8. Files created by CORBA's IDL compiler.

The skeleton files reside in the server and are used to help the implementation of the behavior defined in the interface. One of these classes is *_SoilCORBAInterfaceImplBase*, which will enable our *SoilCORBA* class to provide server-like capabilities. This class inherits part of its behavior from a standard class referred to as *Skeleton*, which implements our interface *SoilCORBAInterface*, as shown in Figure 11-6. Figure 11-9 shows the implementation of class SoilCORBA in Java.

```
1     public class SoilCORBA extends _SoilCORBAInterfaceImplBase {
2         SoilInterface soil = new Soil();
3         public double getSoilDepth(){
4             return soil.getSoilDepth();
5         }
6         public double getWiltingPoint(){
7             return soil.getWiltingPoint();
8         }
9         public void initialize(){
10            soil.initialize();
11        }
12        static void main(String args[]){
```

Figure 11-9. Java implementation of class SoilCORBA (Part 1 of 2).

```
13        try {
14        System.out.println("Starting SoilCORBA Server");
15            org.omg.CORBA.ORB orb =
          org.omg.CORBA.ORB.init(args,null);
16            org.omg.CORBA.BOA boa =
          ((com.visigenic.vbroker.orb.ORB)orb).BOA_init();
17            SoilCORBA soilServer = new SoilCORBA();
18            boa.obj_is_ready(soilServer);
19            System.out.println("Registering server");
20            com.visigenic.vbroker.URLNaming.Resolver resolver =
21               com.visigenic.vbroker.URLNaming.ResolverHelper.narrow(
22                  orb.resolve_initial_references("URLNamingResolver"));
23            String ior = "http://128.227.103.134:1015/SoilServer.ior";
24            resolver.force_register_url(ior, soilServer);
25            soilServer.initialize();
26            boa.impl_is_ready();
27        }
28        catch (Exception e) {
29            System.out.println("CORBA server is not running, check the
          server");
30        }
31    }
32 }
```

Figure 11-9. Java implementation of class SoilCORBA (Part 2 of 2).

Classes *SoilCorba* and *Soil* provide similar behavior as their interfaces they implement are similar. *SoilInterface* and *SoilCORBAInterface* define similar but yet not identical, behavior. What makes the behaviors defined by these interfaces different is the environment in which the corresponding classes are implemented. *SoilInterface* is defined to be implemented by classes that reside in a local computing environment. *SoilCORBAInterface* is defined to be implemented by classes residing in a distributed computing environment. Classes that implement *SoilInterface* need only to provide the behavior defined in the interface. Classes implementing *SoilCORBAInterface*, besides the behavior defined by interface, need to provide server-like capabilities, as they make available their services from a remote server where they reside.

Line 1 in Figure 11-9 defines the class SoilCORBA as inheriting from class *_SoilCORBAInterfaceImplBase,* generated by the IDL compiler. Line 2 creates an instance of class *Soil* to whom class *SoilCORBA* will delegate the receiving method calls for execution. Lines 3 through 11 show the implementation of methods defined in *SoilCORBAInterface*. Note that each of the methods delegates the receiving call for execution to object *soil*. Therefore, class *SoilCORBA* reuses the behavior defined in class *Soil*. Lines 12 through 30 define the static method *main*. This method initializes the environment of the class and must create an instance of the class itself. Line 15 obtains a reference to the ORB and line 16 obtains a reference to a BOA object. Line 17 creates an instance of the class *SoilCORBA* that plays a two-fold role in this application: First, it will provide the behavior needed in the simulation process and second, the behavior of a server. Line 18 notifies the BOA about the existence of object *soilServer* created in line 17; object *soilServer* is passed as a parameter to the BOA. Line 26 notifies that the BOA is ready to receive requests.

Distributed objects are identified by object references that are known as IOR (Interoperable Object Reference). Lines 20 through 24 associate the URL with object's OIR. Once a URL has been bound to an object, client applications can obtain a reference to the object by specifying the URL as a string. The URL Naming Service is the mechanism that lets a server object associate its IOR with a URL in the form of a string in a file. The IOR is composed of server's IP number, a communication port number, and the name assigned to the file. Client programs can then locate the object using the URL pointing to the file on the Web server. Lines 20 through 22 obtain the *Resolver* that will be used to register objects. Line 23 creates a string of the association URL-IOR. Object servers register objects by binding to the Resolver and then using the *force_register_url* method to associate a URL with an object's IOR. Line 25 populates object *soil* with its initial data. Line 26 shows that the server enters a dispatch loop and waits for incoming invocations.

2.5   A simple CORBA client

In order to access the services offered by the server SoilCORBA, a CORBA client needs to be developed. In Java, CORBA clients can be implemented as applets or as applications. Although both methods provide similar results, they are implemented in different ways. We will implement our CORBA client as an application. Figure 11-10 shows the Java implementation of a simple CORBA client. The task of this client is to connect to the *SoilCORBA* server and use one of its services. The service used provides the current value of soil depth parameter.

```
1       import java.util.Properties;
2       public class CORBAClient {
3         public CORBAClient() {}
4         public static void main(String[] args) {
5           //create an instance of the client
6           CORBAClient corbaClient = new CORBAClient();
7           Properties soilProp = new Properties();
8           soilProp.put("org.omg.CORBA.ORBClass",
        "com.visigenic.vbroker.orb.ORB");
9              org.omg.CORBA.ORB orb = org.omg.CORBA.ORB.init((String
        [])null,soilProp);
10             //connect to SoilCORBA server
11         SoilCORBAInterface soil=
12             SoilCORBAInterfaceHelper.bind(orb,
13             "http://128.227.103.134:1015/SoilServer.ior");
14             //print value of soil depth
15         System.out.println("Soil depth value is: "+soil.getSoilDepth());
16       }
17     }
```

Figure 11-10. The implementation of a simple CORBA client in Java.

Line 1 makes available the behavior of class Properties that is needed in lines 7 and 8. Line 2 defines the class *CORBAClient*, and line 3 defines a default constructor for the class. Lines 4 through 16 define the method *main* that initializes the class environment and creates an instance of the class itself in line 6. An object of type *Properties* is created in line 7 to hold the information that Visigenic CORBA technology is used in this case. Line 9 initializes the ORB. Lines 11, 12, and 13 bind the ORB to the IOR of the object located in the server. Therefore, a proxy of the server object is created (see Section 4.2, **The Proxy pattern,** in Chapter 7) in the client application and the communication bridge between server and client is established. The proxy object is object *soil*. Note that the server's IP (Internet Protocol) number and the number of the port used are the same as the ones defined in the server application. If the client and server applications use different IP and/or port numbers, then the communication bridge cannot be established and therefore the client and the server will not be able to communicate with one another. Proxy *soil* will make available to the client application the behavior of the server object. Line 15 uses the proxy to obtain the value of soil depth, which is provided as a service by the server object.

3.   THE REMOTE METHOD INVOCATION (RMI)

While the RMI technology is similar to CORBA, it operates only in the Java programming environment. RMI enables the programmer to develop distributed applications based on methods of Java objects that can be invoked from other Java virtual machines, which can be located on different computers and maybe different operating systems (http://java.sun.com/products/rmi/). RMI provides a simple and direct model for developing distributed applications using Java objects. An RMI client can make calls to an RMI server object and use its behavior as if it were local to it. The RMI technology is a sophisticated mechanism that allows Java objects to communicate amongst them. The mechanism and the communication protocol are well-defined and standardized. A Java object can communicate with another Java object without knowing before hand the protocol used.

During the remote communication between two Java objects, the one that makes the remote call is referred to as the *client object* and the one that responds to the remote call is referred to as the *server object*. Note that the relationship client-server is valid only for this particular call. An object that plays the role of the *client* in one call may be playing the role of the *server* in another call.

RMI uses several layers to achieve the communication between remote objects. These layers are:

1. Client's stubs and server's skeletons; these objects are used to hide the remoteness and make transparent the communication between objects over a network connection.
2. The remote reference layer that handles the packaging of the method call and its parameters and returns the results of method's execution over the network connection.
3. The transport layer that is the actual network connection linking systems together.

To better understand the collaboration between client and server objects using RMI, we will consider a simple example of collaboration between a server object referred as SoilRMIServer, which provides its services remotely and a client object referred to as SoilClient that invokes a method call on the remote server. The services offered by object SoilRMIServer are the same ones that object CORBASoilServer, presented in the previous section, provides. Implementing the same example in two different technologies allows us to better understand the similarities and differences between these technologies.

The stubs and skeletons are automatically generated by the RMI compiler when compiling the *server* class. Figure 11-11 shows the stub and skeleton classes that are generated for class *SoilRMI*.

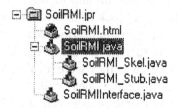

Figure 11-11. Stubs and skeletons created by Java compiler for class SoilRMI.

Stub classes reside with the client and skeletons reside with the server. Stubs and skeletons are involved when a client invokes a remote method on the remote server. The stub object serves as a proxy (see Section 4.2, **The Proxy pattern,** in Chapter 7) for the server and resides with the client. The client's call goes to the stub object that has encapsulated all the names of the methods provided by the server object, as shown in Figure 11-12. Lines 3 through 6 show the names of methods implemented by the server object.

```
1     public final class SoilRMIServer_Stub extends
          java.rmi.server.RemoteStub implements SoilRMIInterface,
          java.rmi.Remote {
2     private static final java.rmi.server.Operation[] operations = {
3         new java.rmi.server.Operation("double getSoilDepth()"),
4         new java.rmi.server.Operation("double getWiltingPoint()"),
5         new java.rmi.server.Operation("void initialize()")
6     };
```

Figure 11-12. Partial Java code for class SoilRMI_Stub.

After receiving the remote call, the stub creates a block of information containing an identifier of the remote object to be used, a description of the method to be called and the marshalled parameters of the remote method. This block of information is sent to the server object using the network connection. On the server side, this information is received by the skeleton which is a server-side object that contains a method that dispatches calls to the actual server object implementation. Skeletons communicate with clients using the JRMP (Java Remote Method Protocol) protocol.

When the skeleton object receives the block of information sent by the stub object, it unmarshals the parameters of the remote call. Then, using the object identifier sent by the stub object, it identifies the remote object responsible for executing the remote call and invokes the method. The result of the remote call is packaged in a block of information and it is sent back to the stub object. Although the communication between client and server objects is complex, it is transparent to the programmer. As previously mentioned, remote objects are used as they were local objects; even the syntax of a remote call is the same as for a local call. Figure 11-13 shows the sequence diagram for the collaboration between the client and the server objects.

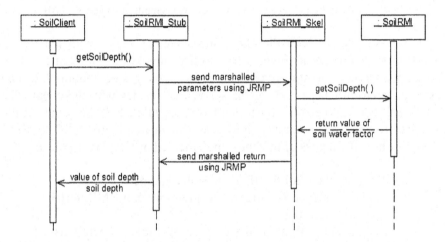

Figure 11-13. Collaboration between client and server objects using RMI.

As shown in Figure 11-13, object *SoilClient* invokes the message *getSoilDepth()* on the remote server object *SoilRMI*. First, the message is sent to the stub object created by the RMI compiler. The stub object creates a block of information and sends it to server object using the JRMP protocol. This block of information is received by the skeleton object that unmarshals it and invokes the method call on the server object. The return of the method call, the value of soil depth, is packaged again in a block of information and is sent back to the stub object by the skeleton object. The stub object unmarshals the return value and sends it to the client object. In the next two sections, we will show a simple example of implementation of a *SoilRMI* server and a *SoilClient*.

3.1   An RMI Soil Server

Let us consider the Soil component presented in the previous section, and implement it as a server using the RMI technology. The *Soil* component will provide its services to any other component or system residing in the same address space or in other servers located anywhere in the network. Therefore, the behavior of component *Soil* will be slightly different; besides the behavior needed in the simulation process, the *Soil* component will be provided with additional behavior to function as an RMI server.

First, an interface needs to be created that will define the functionalities provided by the object server. The interface is referred to as *SoilRMIInterface* and its definition in Java is shown in Figure 11-14. The behavior defined in the interface is the same as the one defined in Section 2.4, **A CORBA Soil Server**, with the only difference that RMI requires that methods defined in the interface should throw an exception, the RemoteException one.

```
1    import java.rmi.RemoteException;
2    import java.rmi.Remote;
3    interface SoilRMIInterface extends Remote {
4        public double getSoilDepth() throws RemoteException;
5        public double getWiltingPoint() throws RemoteException;
6        public void initialize() throws RemoteException;
7    }
```

Figure 11-14. Definition of interface SoilRMIInterface in Java.

As the functionalities defined in the interface will be provided as services that can be invoked remotely, *SoilRMIInterface* should inherit from *Remote* interface, which is part of Java's arsenal for sending/receiving messages over a network. As *SoilRMI* object will provide its services from a remote server, it should be able to instantiate itself with server-like capabilities. Therefore, *SoilRMI* object should inherit the server-like behavior from a predefined Java server class named *UnicastRemoteObject*. The class diagram shown in Figure 11-15 presents links between objects and interfaces for remote communication.

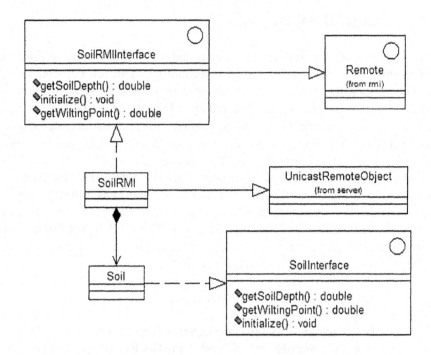

Figure 11-15. Class diagram for the implementation of SoilRMI server.

Class *SoilRMI* implements interface *SoilRMIInterface*; therefore, it will provide an implementation for the methods defined in interface. The behavior of class *SoilRMI* is very similar to the behavior of Soil class, as they play the same role in the simulation process. Therefore, the behavior of class *Soil* can be reused by having class *Soil* be part of *SoilRMI* class definition. *SoilRMI* would be able to provide the functionality required for the simulation by delegating simulation-related messages to the class *Soil*. The behavior required in the simulation process is defined in interface *SoilInterface*. Thus, class *Soil* implements interface *SoilInterface*. The Java implementation for class *SoilRMI* is shown in Figure 11-16.

```
1        import java.rmi.server.*;
2        import java.rmi.Naming;
3        import java.net.*;
4        import java.rmi.registry.*;
5        public class SoilRMIServer extends UnicastRemoteObject
                        implements   SoilRMIInterface {
```

Figure 11-16. Implementation in Java of class SoilRMI (Part 1 of 2).

```
6       public SoilInterface soil = new Soil();
7       public double getSoilDepth(){

8       return soil.getSoilDepth();
9       }
10      public double getWiltingPoint(){
11          return soil.getWiltingPoint();
12      }
13        public void initialize(){
14          soil.initialize();
15      }
16      public SoilRMIServer() throws Exception {}
17      static void main(String [] args) {
18        SoilRMIServer soilServer = null;
19        try {
20          System.out.println("Starting SoilRMI Server");
21          int port = 1012;
22          String name = new String("//128.227.103.134:"+port+"/Soil");
23          soilServer = new SoilRMIServer();
24          LocateRegistry.createRegistry(port);
25          Naming.rebind(name,soilServer);
26          System.out.println("Soil RMI server is running in "+name);
27        }
28        catch(Exception e){System.out.println("Error in  starting  soil
                server");}
29      }
30    }
```

Figure 11-16. Implementation in Java of class SoilRMI (Part 2 of 2).

Lines 1 through 4 show the Java standard libraries that are needed to support the functionalities defined in class *SoilRMI*. Line 5 defines class *SoilRMI* as a subclass of *UnicastRemoteObject* and implementing interface *SoilRMIInterface*. Line 6 defines an attribute of type *Soil* and assigns to it a reference to an object of the same type. This is the implementation in Java of the whole-part relationship between *SoilRMI* and *Soil* as defined in Figure 11-15. Lines 7 through 15 are the implementation of the methods defined in interface *SoilRMIInterface*. Note that each of the methods delegates the call to the object of type *Soil* defined in line 6.

Line 16 defines the constructor for class *SoilRMI*. Lines 17 through 30 define the method *main* that initializes the environment of class *SoilRMI*. Line

18 defines an instance of the class itself to be used later in line 23. Lines 19 through 29 define a **try-catch** block. In the case that an exception occurs while executing lines 19 through 27 in the **try** block, the control goes to the **catch** block and the system prints an error message. Within the **try** block, line 21 defines the communication port and assigns to it a value. Line 22 defines a stringified reference for the server object, including the communication port. Line 23 creates an instance of the class itself. Line 24 creates a registry for the communication port and line 25 binds the stringified reference for the server object to its name created in line 23. Lines 28 and 29 define the **catch** block that contains the actions that need to be executed in the case that an exception occurs within the **try** block.

3.2   A Simple RMI client

In order to access the services offered by the server *SoilRMI*, an RMI client needs to be developed. Figure 11-17 shows the implementation in Java of an RMI client.

```
1        import java.rmi.Naming;
2        import java.rmi.RemoteException;
3        public class RMIClient {
4          void main(String args[]) {
5            //contact remote soil
6            int port = 1012;
7            String url = "rmi://128.227.103.134";
8            url += ":"+port+"/Soil";
9            String registry = new String(url);
10           try {
11             SoilRMIInterface soil = new SoilRMIServer();
12             soil = (SoilRMIInterface)Naming.lookup(registry);
13             double soilDepth = soil.getSoilDepth();
14             System.out.println("RMI Soil Server is available and
                 running");
15           }
16           catch(Exception ex) {
17             System.out.println("RMI Soil Server is not available!");
18           }
19         }
20       }
```

Figure 11-17. Implementation of an RMI client.

Lines 1 and 2 show the Java standard the libraries needed to support the functionalities defined in the class *RMIClient* defined in line 4. Lines 7 through 10 collect information about the server: Its port number and IP (Internet Protocol) address, its stringified reference, and registry. In the case that the IP address and/or the port number are not the ones used by the server, the communication with the server will not be established. Lines 12 through 17 establish contact with the server object to obtain from it the value of soil depth. In the case that the communication with the server is not established, an error message is displayed. Note that the syntax for a remote method is the same as for a local method.

4. DISTRIBUTED CROP SIMULATION MODEL

In Section 4.1, **Conceptual model for the Kraalingen approach,** of Chapter 8, we presented all the classes involved in the simulation such as *Soil, Weather*, and *Plant*, linked by corresponding associations. In the case where a local computing architecture is selected, all the objects will reside in the same address space in the same machine. If a distributed computing architecture is selected, objects may reside in different address spaces and even in different platforms. The fact that objects are located in different machines and platforms does not affect their already conceived behavior. Distributed objects will communicate through a middleware component that in our example is Java's RMI. The middleware provides a common set of management services made available to all classes that have agreed to use this infrastructure, regardless of their location [Bro99]. The process of developing a distributed component-based system can be considered as a three-phase process. In the first phase, a stand-alone application using a local computing architecture is developed. During this step, interfaces and the behavior of the objects/components are designed and carefully tested. Focus is on providing objects with the right behavior and the right interfaces. The entire simulation-based application is tested in order to make sure that the system delivers correct results. In the second phase, an implementation middleware is selected to implement the distributed application. Issues such as how objects will interact remotely and how they would be instantiated are addressed. The third phase deals with issues of developing and implementing objects that would interact remotely, using the selected middleware.

The three-phase layered approach was used to develop the distributed crop simulation model. Issues concerning an object-oriented approach to crop simulation modeling were addressed in developing a stand-alone model programmed in Java [PBJ01]. During this phase, the main focus was on providing flexible objects with the required behavior so they can be reused in the future. Results taken from the object-oriented approach were compared to the ones provided by the existing FORTRAN code presented by [PBJ99]. After successfully finishing the first phase, the next step was selecting the middleware environment needed for developing the distributed application. As the programming language used in the first phase was Java, the following options were available: Using OMG's CORBA or Java's Remote Method Invocation technologies. Any of these technologies could be used as they are conceptually very similar. We chose Java's RMI, as the rest of the application was developed using the Java programming language.

Figure 11-18 shows the interaction among the elements involved in the simulation in a very general manner, without considering any programming language or other implementation details. Defined interfaces describe the messages objects should respond to in order to perform the simulation. Each of the interfaces represents some functionality that is provided by a class/component. To illustrate issues that need to be addressed during the component development of our system, we will focus only on the *Plant* component, as the development of others is carried out in the same manner.

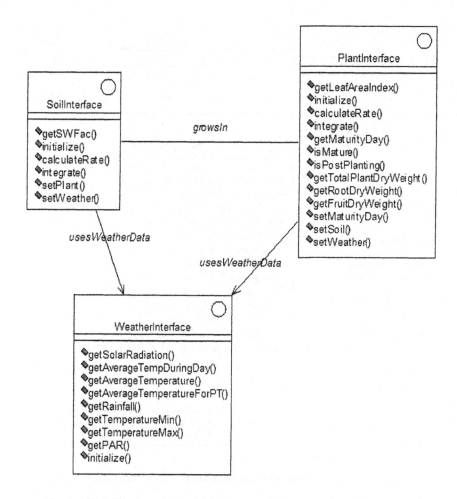

Figure 11-18. Conceptual diagram for the Kraalingen crop simulation model.

As the selected architecture is distributed, the core objects (*Soil, Plant,* and *Weather*) will communicate with each other remotely. Remote communication requires a middleware component that will send messages from one object to another. Figure 11-19 shows the links between objects and interfaces of component *PlantRMI* needed for the remote communication.

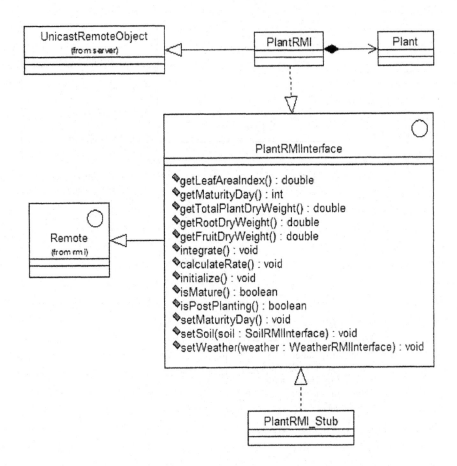

Figure 11-19. Links between objects and interfaces for remote communication.

The interface *PlantRMIInterface* is similar to the *PlantInterface* defined in the conceptual diagram, with the only difference being that it provides capabilities for sending/receiving messages through the Internet. Therefore, *PlantRMIInterface* should inherit the predefined *Remote* interface that is part of Java's arsenal for moving objects in a network. The *PlantRMI* object is functionally similar to the *Plant* object defined in the standalone developed model, as it will play the same role in the simulation process. In addition, *PlantRMI* object should be provided with capabilities of a remote object to be able to remotely communicate with other objects. The behavior of the *Plant* object was defined by *PlantInterface* and the behavior of the object *PlantRMI* is defined by *PlantRMIInterface. PlantRMI* will be able to provide the functionality required for the simulation by delegating simulation-related messages to object *Plant*. *Plant* is part of object *PlantRMI* [GHJ95], [Gra98]; the association linking objects *PlantRMI* and *Plant* is a composition. Because *PlantRMIInterface* inherits from *Remote* interface, the *PlantRMI* object will provide the capabilities for sending/receiving messages over a network. In addition, as the *PlantRMI* object will provide its services from a remote server, it should be able to instantiate itself with server-like capabilities. Therefore, the *PlantRMI* object should inherit the server-like behavior from a predefined Java server class named *UnicastRemoteObject*. Figure 11-20 shows the entire class diagram for the application.

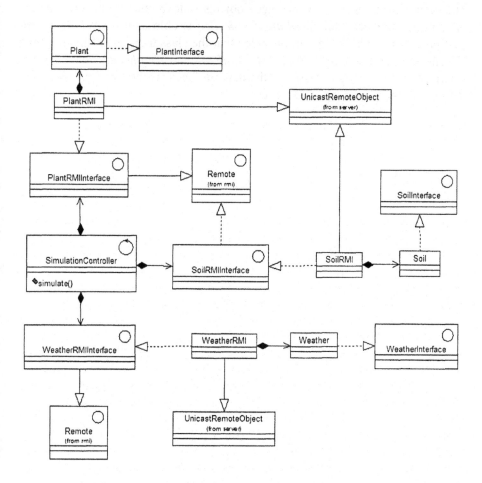

Figure 11-20. Class diagram for Kraalingen distributed system.

As shown in Figure 11-21, the center of the class diagram is the *SimulationController* class. This is a controller type of class that will coordinate the communication between entity classes, by sending them the right message at the right time. In addition, *SimulationController* will supervise the communication with the boundary class. In Figure 11-20, the communication with boundary class is not shown. *SimulationController* has access to interfaces *PlantRMIInterface*, *SoilRMIInterface*, and *WeatherRMIInterface*. The association between the control class and interfaces is modeled as a composition; *SimulationController* being the "whole" and the interfaces being the "parts." Therefore, *SimulationController* is responsible for creating instances of classes implementing these interfaces and controlling the flow of messages objects sent to one another. By having access to interfaces, the *SimulationController* is able to use any other class that provides similar behavior, provided that the class implements the required interface. Note that classes *PlantRMI*, *SoilRMI*, and *WeatherRMI* have similar relationships with other classes of the diagram because they will provide their services in the same way, as a server.

EPILOGUE

Although we have included in this book many chapters dealing with several interesting subjects, we are conscious that there are others that are left out. Topics such as modeling Web-based or XML-based applications are not treated in this book. Future publications need to address these important topics.

At the time this book was published a new modeling paradigm just surfaced in the software engineering world: The Model Driven Architecture (MDA) approach. Few MDA-based tools were available, such Optimal-J of Compuware (http://www.compuware.com/), OlivaNova of SosyInc (http://sosyinc.com/), Project Technology, Inc. (http://www.projtech.com), Kennedy-Carter, Ltd (http://www.kc.com/), Kabira Technologies, Inc. (http://www.kabira.com/), and others.

This new technology looks very promising and it is a UML-based technology. MDA uses UML almost as a programming language. We hope that our book paves the way to the next challenge: The use of the MDA approach to model agricultural systems.

GLOSSARY

abstract class   A class that cannot be instantiated and can be used
                    only as a superclass of other classes.

abstraction   The essential characteristics of an entity that make it
                    different from other entities.

activity diagram   A diagram that shows the flow from activity to
                    activity; activity diagrams address the dynamic view
                    of a system. A special case of a state diagram in
                    which all or most of the states are activity states and
                    in which all or most of the transitions are triggered by
                    completion of activities in the source states [BRJ99].

actor   A coherent set of roles that users of use cases play
                    when interacting with the use cases [BRJ99].

aggregation   A specific type of association used to represent the
                    whole-part relationships.

architecture   The set of significant decisions about the organization
                    of a software system, the selection of the structural
                    elements and their interfaces by which the system is
                    composed, together with their behavior as specified in
                    the collaboration among those elements, the
                    composition of these structural and behavioral
                    elements into progressively larger subsystems, and

the architectural style that guides this organization– these elements and their interfaces, their collaborations, and their compositions. Software architecture is not only concerned with structure and behavior, but also with usage, functionality, performance, resilience, reuse, comprehensibility, economic and technology constraints and trade-offs, and aesthetic concerns [BRJ99].

association   A structural relationship that describes a set of links, in which a link is a connection among objects; the semantic relationship between two or more classifiers that involves the connection among their instances [BRJ99].

attribute   A named property of a classifier that describes a range of values that instances of the property may hold [BRJ99].

behavior   The observable effects of an event, including its results [BRJ99].

binary association   An association between two classes [BRJ99].

cardinality   The number of elements in a set.

class   A description of a set of objects that share the same attributes, operations, relationships, and semantics [BRJ99].

class diagram   A diagram that represents a number of classes, interfaces, and their relationships.

classifier   A mechanism that describes structural and behavioral features. Classifiers include classes, interfaces, datatypes, signals, components, nodes, use cases, and subsystems [BRJ99].

collaboration diagram
> A diagram that represents the collaboration between classes; shows how classes are interrelated and the messages objects created from these classes send and receive.

component
> A physical and replaceable part of a system that conforms to and provides the realization of a set of interfaces [BRJ99].

component diagram
> A diagram that shows the organization of and dependencies among a set of components; component diagrams address the static implementation view of a system [BRJ99].

composition
> A form of aggregation with strong ownership and coincident lifetime of the parts by the whole; parts with nonfixed multiplicity may be created after the composite itself, but once created, they live and die with it; such parts can also be explicitly removed before the death of the composite [BRJ99].

concrete class
> A class that can create instances.

constraint
> A restriction on one or more values of (part of) an object oriented model or system [WK99].

container
> An object created to contain other objects and that provides facilities to access or iterate over them.

delegation
> The ability of an object to defer the execution of a call to another object.

diagram
> The graphical presentation of a set of elements, most often rendered as a connected graph of vertices (things) and arcs (relationships) [BRJ99].

domain
> An area of knowledge or activity characterized by a set of concepts and terminology understood by practitioners in the area [BRJ99].

element   An atomic constituent of a model [BRJ99].

focus of control   A symbol on a sequence diagram that shows the period of time during which an object is performing an action directly or through a subordinate operation [BRJ99].

inheritance   The mechanism by which more-specific elements incorporate the structure and behavior of more-general elements [BRJ99].

instance   A concrete manifestation of an abstraction [BRJ99].

interaction diagram   A diagram that shows an interaction, consisting of a set of objects and their relationships, including the messages that may be dispatched among them; interaction diagrams address the dynamic view of a system; a generic term that applies to several types of diagrams that emphasize object interaction, including collaboration diagrams, sequence diagrams, and activity diagrams [BRJ99].

interface   a set of operations that are used to define the behavior of a class or component.

link   A semantic connection among objects; an instance of an association [BRJ99].

method   The implementation of an operation [BRJ99].

model   A simplification of reality, created in order to better understand the system being created; a semantically closed abstraction of a system [BRJ99].

multiple inheritance   A semantic variation of generalization in which a child may have more than one parent [BRJ99].

multiplicity   A specification of the range of allowable cardinalities that a set may assume [BRJ99].

object   A synonym for an instance.

object constraint language (OCL)
        A formal language used to express constraints over elements of a model.

object diagram
        A diagram that shows a set of objects and their relationships at a point in time; object diagrams address the static design view or static process view of a system [BRJ99].

object lifeline
        A line in a sequence diagram that represents the existence of an object over a period of time [BRJ99].

package
        A general-purpose mechanism for organizing elements into groups [BRJ99].

pattern
        A pattern names, abstracts, and identifies the key aspects of a common design structure that make it useful for creating a reusable object-oriented design [GHJ95].

postcondition
        A constraint that must be true at the completion of an operation [BRJ99].

precondition
        A constraint that must be true when an operation is invoked [BRJ99].

property
        A named value denoting a characteristic of an element [BRJ99].

realization
        A semantic relationship between classifiers, in which one classifier specifies a contract that another classifier guarantees to carry out [BRJ99].

relationship
        A semantic connection among elements [BRJ99].

sequence diagram
        An interaction diagram that emphasizes the time ordering of messages [BRJ99].

single inheritance
        A semantic variation of generalization in which a child may have only one parent [BRJ99].

state   A condition or situation during the life of an object during which it satisfies some condition, performs some activity, or waits for some event [BRJ99].

statechart diagram   A diagram that shows a state machine; statechart diagrams address the dynamic view of a system [BRJ99].

state machine   A behavior that specifies the sequences of states an object goes through its lifetime in response to events, together with its responses to those events [BRJ99].

UML   The Unified Modeling Language, a language for visualizing, specifying, constructing, and documenting the artifacts of a software-intensive system [BRJ99].

use case   A description of a set of actions, including variants, that a system performs that yields an observable result of value to an actor [BRJ99].

use case diagram   A diagram that shows a set of use cases and actors and their relationships; use case diagrams address the static use case view of a system [BRJ99].

use case model   The set of all use cases for a problem represented in one diagram.

REFERENCES

[AR97]   Acock, B., Reddy, V.R.: Designing an object-oriented structure for crop models. Ecological Modelling 94, 33-44 (1997)

[Alex77] Alexander, C. A.: Pattern Language: Towns, Buildings, Constructions, Oxford University Press, New York (1977)

[Alex79] Alexander, C.: The Timeless Way of Building. Oxford University Press, Oxford, UK (1979)

[AW85]   Addiscott, T.M.: Wagenet, R.J. Concepts of solute leaching in soils: a review of modeling approaches. Journal of Soil Science 36, 411–424 (1985)

[BJ98]   Beck, H., Jackson, J.: Florida automated weather network. In: Proceedings of the Seventh International Conference on Computers in Agriculture, Orlando, Florida, ASAE, St. Joseph, MI, pp. 595–601 (1998)

[BC87]   Beck, K. and Cunningham W.: "Window per Task" http://c2.comcgi/wiki? WindowPerTask (1987)

[BBC99] Bernstein, P.A, Bergstraesser, T., *Carlson*, J.,   Pal S., Sanders, P., Shutt, D.: Microsoft repository version 2 and the open information model. Information Systems 24(2) 71-98 (1999)

[BB02]   Boggs, W., Boggs, M.: UML with Rational Rose 2002 SYBEX (2002)

[Boo86] Booch, G.: Object-Oriented development. IEEE Transactions on Software Engineering 12 (2), p.211-221 (1986)

[Boo91] Booch, G.: Object Oriented design with Applications. Benjamin/Cummings (1991)

[BRJ99] Booch, G., Rumbaugh, J., Jacobson, I.: The Unified Modeling Language User Guide. Addison-Wesley, Reading, MA (1999)

[BKT00] Bouman, B.A.M., Kropff, M.J., Tuong, T.P., Wopereis, M.C.S., von Berge, H.H.M., van Laar H.H.: ORYZA2000: modeling lowland rice, International Rice Research Institute (IRRI) Metro Manila, Philippines (2000)

[BC90] Bracha, G., Cook, W.: Mixin-based inheritance. In OOPSLA/ECOOP'90 Conference Proceedings (Ottawa, Canada, Oct. 21-25). ACM SIGPLAN Not. 25, 10 (Oct.) 303-311 (1990)

[BS03] Bittner, K., Spencer, I.: Use Case Modeling. Addison-Wesley (2003)

[Bro99] Brown A.: Constructing Superior Software. Macmillan Technical Publishing, USA (1999)

[BW98] Brown, A., Wallnau, K.C.: The Current State of CBSE. Sterling Software. Software Engineering Institute. September/ October. IEEE Software. 15, No. 5, 37-46 (1998)

[Bar03] Business Component-Based Software Engineering, Editor: Barbier F, Kluwer Academic Publishers, 2003

[CMK99] Coad, P., Mayfield, M., Kern, J.: Java Design, Building Better Apps & Applets. Yourdon Press Computing Series. Prentice-Hall, (1999)

[CY91] Coad, P. and Yourdon, E.: Object-Oriented Analysis. Prentice Hall, Englewood Cliffs (1991)

[CAB94] Coleman, D., Arnold, P., Bodoff, S., Dollin, C., Gilchrist, H., Hayes, F. Jeremaes, P.: Object-Oriented Development - The Fusion Method, Prentice-Hall Englewood Cliffs, N.J. (1994)

[C00] Conallen J.: Building Web Applications With UML. Addison Wesley (2000)

[Coo87] Cook, S.: OOPSLA Panel p 2: Varieties of inheritance. In OOPSLA'87 Addendum to the Proceedings, pp. 35-40. ACM Press October (1987)

[CV90] Chopart, J.L., Vauclin, M. Water balance estimation model: field test and sensitivity analysis. Soil Science Society of America Journal 54, 1377–1384 (1990)

[CJH94] Clemente, R.S., De Jong, R., Hayhoe, H.N., Reynolds, W.D., Hares, M.: Testing and comparison of three unsaturated soil water flow models. Agricultural Water Management 25, 135–152 (1994)

[CK87] Cunningham, W., Beck, K.: Using Pattern Languages for Object-oriented Programs. OOPSLA Conference (1987)

[DMN68] Dahl, O-J., Myhrhaug, B., Nygaard, K.: SIMULA 68 common base language. Tech. Rep., Norwegian Computing Center, Oslo, May (1968)

[DP01] Drouet, J.-L., Pages, L.: Object-oriented modeling of the relationships between growth and assimilates partitioning from the organ to the whole plant. Second International Symposium Modelling Cropping Systems, Firenze, Italy, July 16–18, pp.19–20 (2001)

[Eag78] Eagleson, P.S.: Climate, soil and vegetation. A simplified model of soil moisture movement in the liquid phase. Water Resources Research 14, 722–730 (1978)

[Fra03] Frankel, S. D.: Model Driven Architecture. Applying MDA to Enterprise Computing. Wiley Publishing, Inc. OMG Press (2003)

[GHJ95] Gamma, E., Helm, R., Johnson, R., Vlissides, J.: Design Patterns Elements of Reusable Object-Oriented Software. Addison-Wesley, Reading, MA (1995)

[GSR00] George, B.A., Shende, S.A., Raghuwanshi, N.S.: Development and testing of an irrigation scheduling model. Agricultural Water Management 46, 121–136 (2000)

[Gra98] Grand, M.: Patterns in Java. Vol 1. Wiley Computer Publishing. John Wiley & Sons, Inc (1998)

[Hea94] Hearn, A.B.: OZCOT, a simulation model for cotton crop management. Agricultural Systems 44, 257–299 (1994)

[HSL01] Huang, M., Shao, M., Li, Y.: Comparison of a modified statistical-dynamic water balance model with the numerical model WAVES and field measurements. Agricultural Water Management 48, 21–35 (2001)

[HWH01] Hunt, L.A., White, J.W., Hoogenboom, G.:Agronomic data: advances in documentation and protocols for exchange and use. Agricultural Systems 70 477–492, 2001

[JBR98] Jacobson, I., Booch, G., Rumbaugh, J.: The Unified Software Development Process. Addison-Wesley, Reading, MA (1998)

[JCJ94] Jacobson, I., Christerson, M., Jonsson G. O.: Object-oriented Software Engineering. A Use Case Driven Approach. Addison-Wesley (1994)

[Kra95] D.W.G. van Kraalingen.: The FSE system for crop simulation, version 2.1. Quantitative Approaches in Systems Analysis Report no. 1. AB/DLO, PE, Wageningen (1995)

[Kru98] Krutchen, Ph.: The Rational Unified Process, An Introduction. Addison-Wesley (1998)

[Lak96] Lakos, J.: Large Scale C++ Software Design. Addison Wesley (1996)

[Lar02] Larman, K.: Applying UML and patterns: an introduction to object-oriented analysis and design and the Unified Process. Prentice Hall, Inc. (2002)

[LKN02] Laurenson, M., Kiura, T., Ninomiya, S.: Providing agricultural models with mediated access to heterogeneous weather databases. Applied Engineering in Agriculture 17, 617–625 (2002)

[LEM01] Leib, B.G., Elliott, T.V., Matthews, G.: WISE: a web-linked and producer oriented program for irrigation scheduling. Computers and Electronics in Agriculture 33, 1–6 (2001)

[LCh97] Lemmon, H., Chuk, N.: Object-oriented design of a cotton crop model. Ecol. Model. 94, 45_ 51 (1997)

[MLB98] Maraux, F., Lafolie, F., Bruckler, L.: Comparison between mechanistic and functional models for estimating soil water balance: deterministic and stochastic approaches. Agricultural Water Management.38, 1–20 (1998)

[MB02] Mellor J. S., Balcer J. M.: Executable UML, A Foundation for Model-Driven Architecture, Addison Wesley (2002)

[Mey88] Meyer, B.: Object-Oriented software construction. Prentice Hall (1988)

[OEK01] Olejnik, J., Eulenstein, F., Kedziora, A., Werner, A.: Evaluation of a water balance model using data for bare soil and crop surfaces in Middle Europe. Agricultural and Forest Meteorology 106, 105–116 (2001)

[Pap05] Papajorgji, P.: A plug and play approach for developing environmental models. Environmental Modelling & Software, Vol 20/10 pp. 1353-1357 (2005)

[PBJ01] Papajorgji, P., Braga, R., Jones, J.W., Porter, C., Beck, H.W.: Object-Oriented design of crop models with interchangeable modules: concepts and issues. 2nd International Symposium, Modelling Cropping Systems, Florence, Italy (2001)

[PSH04] Papajorgji, P., Shatar, T.: Using the Unified Modelling Language to develop soil water-balance and irrigation-scheduling models. Environmental Modelling & Software 19 (2004) 451-459 (2004)

[PBB04] Papajorgji, P., Beck, W.B., Braga, J.L.: An Architecture for developing service-oriented and component-based environmental models. Ecological Modelling, Vol 179/1, pp. 61-76 (2004)

[PR02] Pardalos, P.M. and Resende, M. (Eds), Handbook of Applied Optimization, Oxford University Press (2002)

[PSZ95] Pardalos, P.M., Siskos Y., and Zopounidis C. (Eds), Advances in Multicriteria Analysis, Kluwer Academic Publishers, (1995)

[Ped89] Pedersen, H. C.: Extending Ordinary Inheritance Schemes to Include Generalization. ACM SIGPLAN Not. 24, 10 (Oct.), pp. 407-417 (1989)

[PL82]   Penning de Vries, FWT., van Laar, H.H.   editors. Simulation of plant growth and crop production. Simulation Monographs. Wageningen , Netherlands (1982)

[PBJ99] Porter, C.H., Braga, R., Jones, J.W.: Research Report No 99-0701, Agricultural and Biological Engineering Department, University of Florida, Gainesville, Florida, USA (1999)

[Pro02] Prosise, J.: Programming Microsoft .NET. Microsoft Press (2002)

[Ros98] Rosenberger, Jeremy L.: Teach Yourself CORBA in 14 Days (SAMS) (1998)

[Rit98] Ritchie, J.T.: Soil water balance and plant water stress. In: Tsuji, G.Y., Hoogenboom, G., Thornton, P.K. (Eds.), Understanding Options for Agricultural Production. Kluwer Academic Publishers, Dordrecht (1998)

[RO85] Ritchie, J.T., Otter, S.: Description and performance of CERESwheat: a user-oriented wheat yield model. In: Muchow, R.C., Bellamy, J.A. (Eds.), Climate Risk in Crop Production: Models and Management for the Semiarid Tropics and Subtropics. CAB International, Wallingford, OXON (1985)

[ShM88] Shlaer, S. and Mellor, S. J.: Object-Oriented Systems Analysis - Modeling the World in Data, Prentice-Hall, Englewood Cliffs, N.J. (1988)

[Szy99] Szyperski, C.: Component Software. Beyond Object-Oriented Programming. Addison-Wesley. ACM Press. New York (1999)

[SDN03] Scharli, N., Ducasse, S., Nierstrasz, O., and Black, A.: Traits: Composable Units of Behavior. In OOPSLA/ECOOP' 2003 Conference Proceedings (2003)

[Tai96] Taivalsaari, A.: On the notion of inheritance. ACM Computing Surveys, vol 28, No. 3 (1996)

[VD98] Vogel, A., Duddy, K.: JAVA Programming with CORBA. Wiley Computer Publishing (1998)

[Vin97] Vinovski S.: CORBA: Integrating Diverse Applications Within Distributed Heterogeneous Environments. IEEE Communications Magazine, Vol. 35, No. 2 (1997)

[Vog03] Vogels, W.: Web Services are not Distributed Objects: Common Misconceptions about Service Oriented Architectures. http://weblogs.cs.cornell.edu/AllThingsDistributed/archives/000120.html (2003)

[Weg00] Wegehenkel, M.: Test of a modelling system for simulating water balances and plant growth using various different complex approaches. Ecological Modelling 129, pp. 39-64 (2000)

[WWW94] Waldo, J., Wyant, G., Wollrath, A., Kendall, S.:   A Note on Distributed Computing. Sun Microsystems Laboratories, Mountain View, CA 94043 (1994)

[WK99] Warmer, J., Kleppe, A.: The Object Constraint Language, Precise Modeling with UML. Addison Wesley (1999)

[Zdo99] Zdonik, S.: Why properties are objects or some refinements of "is-a". Proceedings of 1986 fall joint computer conference on Fall joint computer conference. Publisher: IEEE Computer Society Press Los Alamitos, CA (1999)

INDEX